Research Notes in Mathematics

Submission of proposals for consideration
Suggestions for publication, in the form of outlines and representative samples, are invited by the Editorial Board for assessment. Intending authors should approach one of the main editors or another member of the Editorial Board, citing the relevant AMS subject classifications. Alternatively, outlines may be sent directly to one of the publisher's offices. Refereeing is by members of the board and other mathematical authorities in the topic concerned, throughout the world.

Preparation of accepted manuscripts
On acceptance of a proposal, the publisher will supply full instructions for the preparation of manuscripts in a form suitable for direct photo-lithographic reproduction. Specially printed grid sheets are provided and a contribution is offered by the publisher towards the cost of typing. Word processor output, subject to the publisher's approval, is also acceptable.

Illustrations should be prepared by the authors, ready for direct reproduction without further improvement. The use of hand-drawn symbols should be avoided wherever possible, in order to maintain maximum clarity of the text.

The publisher will be pleased to give any guidance necessary during the preparation of a typescript, and will be happy to answer any queries.

Important note
In order to avoid later retyping, intending authors are strongly urged not to begin final preparation of a typescript before receiving the publisher's guidelines and special paper. In this way it is hoped to preserve the uniform appearance of the series.

Advanced Publishing Program
Pitman Publishing Inc
1020 Plain Street
Marshfield, MA 02050, USA
(tel (617) 837 1331)

Advanced Publishing Progra
Pitman Publishing Limited
128 Long Acre
London WC2E 9AN, UK
(tel 01-379 7383)

Titles in this series

1 Improperly posed boundary value problems
A Carasso and A P Stone
2 Lie algebras generated by finite dimensional ideals
I N Stewart
3 Bifurcation problems in nonlinear elasticity
R W Dickey
4 Partial differential equations in the complex domain
D L Colton
5 Quasilinear hyperbolic systems and waves
A Jeffrey
6 Solution of boundary value problems by the method of integral operators
D L Colton
7 Taylor expansions and catastrophes
T Poston and I N Stewart
8 Function theoretic methods in differential equations
R P Gilbert and R J Weinacht
9 Differential topology with a view to applications
D R J Chillingworth
10 Characteristic classes of foliations
H V Pittie
11 Stochastic integration and generalized martingales
A U Kussmaul
12 Zeta-functions: An introduction to algebraic geometry
A D Thomas
13 Explicit *a priori* inequalities with applications to boundary value problems
V G Sigillito
14 Nonlinear diffusion
W E Fitzgibbon III and H F Walker
15 Unsolved problems concerning lattice points
J Hammer
16 Edge-colourings of graphs
S Fiorini and R J Wilson
17 Nonlinear analysis and mechanics: Heriot-Watt Symposium Volume I
R J Knops
18 Actions of fine abelian groups
C Kosniowski
19 Closed graph theorems and webbed spaces
M De Wilde
20 Singular perturbation techniques applied to integro-differential equations
H Grabmüller
21 Retarded functional differential equations: A global point of view
S E A Mohammed
22 Multiparameter spectral theory in Hilbert space
B D Sleeman
24 Mathematical modelling techniques
R Aris
25 Singular points of smooth mappings
C G Gibson
26 Nonlinear evolution equations solvable by the spectral transform
F Calogero
27 Nonlinear analysis and mechanics: Heriot-Watt Symposium Volume II
R J Knops
28 Constructive functional analysis
D S Bridges
29 Elongational flows: Aspects of the behaviour of model elasticoviscous fluids
C J S Petrie
30 Nonlinear analysis and mechanics: Heriot-Watt Symposium Volume III
R J Knops
31 Fractional calculus and integral transforms of generalized functions
A C McBride
32 Complex manifold techniques in theoretical physics
D E Lerner and P D Sommers
33 Hilbert's third problem: scissors congruence
C-H Sah
34 Graph theory and combinatorics
R J Wilson
35 The Tricomi equation with applications to the theory of plane transonic flow
A R Manwell
36 Abstract differential equations
S D Zaidman
37 Advances in twistor theory
L P Hughston and R S Ward
38 Operator theory and functional analysis
I Erdelyi
39 Nonlinear analysis and mechanics: Heriot-Watt Symposium Volume IV
R J Knops
40 Singular systems of differential equations
S L Campbell
41 N-dimensional crystallography
R L E Schwarzenberger
42 Nonlinear partial differential equations in physical problems
D Graffi
43 Shifts and periodicity for right invertible operators
D Przeworska-Rolewicz
44 Rings with chain conditions
A W Chatters and C R Hajarnavis
45 Moduli, deformations and classifications of compact complex manifolds
D Sundararaman
46 Nonlinear problems of analysis in geometry and mechanics
M Atteia, D Bancel and I Gumowski
47 Algorithmic methods in optimal control
W A Gruver and E Sachs
48 Abstract Cauchy problems and functional differential equations
F Kappel and W Schappacher
49 Sequence spaces
W H Ruckle
50 Recent contributions to nonlinear partial differential equations
H Berestycki and H Brezis
51 Subnormal operators
J B Conway
52 Wave propagation in viscoelastic media
F Mainardi
53 Nonlinear partial differential equations and their applications: Collège de France Seminar. Volume I
H Brezis and J L Lions
54 Geometry of Coxeter groups
H Hiller
55 Cusps of Gauss mappings
T Banchoff, T Gaffney and C McCrory

56 An approach to algebraic K-theory
A J Berrick
57 Convex analysis and optimization
J-P Aubin and R B Vintner
58 Convex analysis with applications in the differentiation of convex functions
J R Giles
59 Weak and variational methods for moving boundary problems
C M Elliott and J R Ockendon
60 Nonlinear partial differential equations and their applications: Collège de France Seminar. Volume II
H Brezis and J L Lions
61 Singular systems of differential equations II
S L Campbell
62 Rates of convergence in the central limit theorem
Peter Hall
63 Solution of differential equations by means of one-parameter groups
J M Hill
64 Hankel operators on Hilbert space
S C Power
65 Schrödinger-type operators with continuous spectra
M S P Eastham and H Kalf
66 Recent applications of generalized inverses
S L Campbell
67 Riesz and Fredholm theory in Banach algebra
B A Barnes, G J Murphy, M R F Smyth and T T West
68 Evolution equations and their applications
F Kappel and W Schappacher
69 Generalized solutions of Hamilton-Jacobi equations
P L Lions
70 Nonlinear partial differential equations and their applications: Collège de France Seminar. Volume III
H Brezis and J L Lions
71 Spectral theory and wave operators for the Schrödinger equation
A M Berthier
72 Approximation of Hilbert space operators I
D A Herrero
73 Vector valued Nevanlinna Theory
H J W Ziegler
74 Instability, nonexistence and weighted energy methods in fluid dynamics and related theories
B Straughan
75 Local bifurcation and symmetry
A Vanderbauwhede
76 Clifford analysis
F Brackx, R Delanghe and F Sommen
77 Nonlinear equivalence, reduction of PDEs to ODEs and fast convergent numerical methods
E E Rosinger
78 Free boundary problems, theory and applications. Volume I
A Fasano and M Primicerio
79 Free boundary problems, theory and applications. Volume II
A Fasano and M Primicerio
80 Symplectic geometry
A Crumeyrolle and J Grifone
81 An algorithmic analysis of a communication model with retransmission of flawed messages
D M Lucantoni
82 Geometric games and their applications
W H Ruckle
83 Additive groups of rings
S Feigelstock
84 Nonlinear partial differential equations and their applications: Collège de France Seminar. Volume IV
H Brezis and J L Lions
85 Multiplicative functionals on topological algebras
T Husain
86 Hamilton-Jacobi equations in Hilbert spaces
V Barbu and G Da Prato
87 Harmonic maps with symmetry, harmonic morphisms and deformations of metrics
P Baird
88 Similarity solutions of nonlinear partial differential equations
L Dresner
89 Contributions to nonlinear partial differential equations
C Bardos, A Damlamian, J I Díaz and J Hernández
90 Banach and Hilbert spaces of vector-valued functions
J Burbea and P Masani
91 Control and observation of neutral systems
D Salamon
92 Banach bundles, Banach modules and automorphisms of C*-algebras
M J Dupré and R M Gillette
93 Nonlinear partial differential equations and their applications: Collège de France Seminar. Volume V
H Brezis and J L Lions
94 Computer algebra in applied mathematics: an introduction to MACSYMA
R H Rand
95 Advances in nonlinear waves. Volume I
L Debnath
96 FC-groups
M J Tomkinson
97 Topics in relaxation and ellipsoidal methods
M Akgül
98 Analogue of the group algebra for topological semigroups
H Dzinotyiweyi
99 Stochastic functional differential equations
S E A Mohammed
100 Optimal control of variational inequalities
V Barbu
101 Partial differential equations and dynamical systems
W E Fitzgibbon III
102 Approximation of Hilbert space operators. Volume II
C Apostol, L A Fialkow, D A Herrero and D Voiculescu
103 Nondiscrete induction and iterative processes
V Ptak and F-A Potra
104 Analytic functions – growth aspects
O P Juneja and G P Kapoor
105 Theory of Tikhonov regularization for Fredholm equations of the first kind
C W Groetsch

106 Nonlinear partial differential equations and free boundaries. Volume I
J I Díaz
107 Tight and taut immersions of manifolds
T E Cecil and P J Ryan
108 A layering method for viscous, incompressible L_p flows occupying R^n
A Douglis and E B Fabes
109 Nonlinear partial differential equations and their applications: Collège de France Seminar. Volume VI
H Brezis and J L Lions
110 Finite generalized quadrangles
S E Payne and J A Thas
111 Advances in nonlinear waves. Volume II
L Debnath
112 Topics in several complex variables
E Ramírez de Arellano and D Sundararaman
113 Differential equations, flow invariance and applications
N H Pavel
114 Geometrical combinatorics
F C Holroyd and R J Wilson
115 Generators of strongly continuous semigroups
J A van Casteren
116 Growth of algebras and Gelfand–Kirillov dimension
G R Krause and T H Lenagan
117 Theory of bases and cones
P K Kamthan and M Gupta
118 Linear groups and permutations
A R Camina and E A Whelan
119 General Wiener–Hopf factorization methods
F-O Speck
120 Free boundary problems: applications and theory, Volume III
A Bossavit, A Damlamian and M Fremond
121 Free boundary problems: applications and theory, Volume IV
A Bossavit, A Damlamian and M Fremond
122 Nonlinear partial differential equations and their applications: Collège de France Seminar. Volume VII
H Brezis and J. L. Lions
123 Geometric methods in operator algebras
H Araki and E G Effros
124 Infinite dimensional analysis–stochastic processes
S Albeverio
125 Ennio de Giorgi Colloquium
P Krée
126 Almost-periodic functions in abstract spaces
S Zaidman
127 Nonlinear variational problems
A Marino, L Modica, S Spagnolo and M Degiovanni
128 Second-order systems of partial differential equations in the plane
L K Hua, W Lin and C-Q Wu
129 Asymptotics of high-order ordinary differential equations
R B Paris and A D Wood
130 Stochastic differential equations
R Wu

Nonlinear partial differential equations and their applications
Collège de France Seminar
VOLUME VII

H Brezis & J L Lions (Editors)
D Cioranescu (Coordinator)
Université Pierre et Marie Curie (Paris VI)

Nonlinear partial differential equations and their applications Collège de France Seminar VOLUME VII

Pitman Advanced Publishing Program
BOSTON · LONDON · MELBOURNE

PITMAN PUBLISHING INC
1020 Plain Street, Marshfield, Massachusetts 02050

PITMAN PUBLISHING LIMITED
128 Long Acre, London WC2E 9AN

Associated Companies
Pitman Publishing Pty Ltd, Melbourne
Pitman Publishing New Zealand Ltd, Wellington
Copp Clark Pitman, Toronto

First published 1985

AMS Subject Classifications: 35-XX, 34-XX, 46-XX

ISSN 0743-0337

Library of Congress Cataloging in Publication Data
(Revised for volume VII)
Main entry under title:

Nonlinear partial differential equations and their
applications.

(Research notes in mathematics;)
Lectures presented at the weekly Seminar on Applied
Mathematics, Collège de France, Paris.
English and French.
Includes bibliographical references.
1. Differential equations, Partial—Congresses.
2. Differential equations, Nonlinear—Congresses.
I. Brézis, H. (Haim) II. Lions, Jacques Louis.
III. Seminar on Applied Mathematics. IV. Series.

QA377.N67 1982 515.3′53 81-4350
ISBN 0-273-08491-7 (v. 1)
ISBN 0-273-08541-7 (v. 2)

British Library Cataloguing in Publication Data

Nonlinear partial differential equations and
their applications: College de France Seminar.
—(Research notes in mathematics, ISSN 0743-0337; 122)
Vol. 7
1. Differential equations, Partial
2. Differential equations, Nonlinear
I. Brezis, H. II. Lions, J.L.
III. Cioranescu, D. IV. Series
515.3′53 QA377

ISBN 0-273-08679-0

Reproduced and printed by photolithography
in Great Britain by Biddles Ltd, Guildford

Preface

The present volume consists of written versions of lectures held during the year 1983-1984 at the weekly Seminar an Applied Mathematics at the College de France. They mostly deal with various aspects of the theory of non linear partial differential equations.

We thank :

- the speakers who kindly agreed to write up their lectures.
- Mrs Doïna Cioranescu who has coordinated the activities of the Seminar and prepared the material for publication ; without her patience and determination this volume would never have appeared.
- Mrs Force, for her competent typing of the manuscripts.

Paris, January 1985

Haïm BREZIS Jacques Louis LIONS

P.S. The seminar is partially supported by a Grant from the C.N.R.S.

Préface

Ce volume regroupe les textes des conférences données en 1983-1984 au Séminaire de Mathématiques appliquées qui se réunit chaque semaine au Collège de France. Elles concernent principalement l'étude d'équations aux dérivées partielles non linéaires sous des éclairages variés.

Nous remercions vivement :

- les conférenciers qui ont bien voulu accepter de rédiger leurs exposés,
- Mme Doïna Cioranescu qui s'est chargée de coordonner les activités du Séminaire et de la préparation matérielle de cet ouvrage ; sans sa patience et sa persévérance cette publication n'aurait pas vu le jour,
- Mme Force, qui a tapé avec compétence les manuscrits.

Paris, Janvier 1985

Haïm BREZIS Jacques Louis LIONS

P.S. Le séminaire était subventionné en partie par les crédits d'une RCP du C.N.R.S.

Contents

Abstracts – Résumés

A.V. BABIN, M.I. VISHIK. ATTRACTEURS MAXIMAUX DANS LES EQUATIONS AUX DERIVEES PARTIELLES.

Abstract : One considers various systems of partial differential equations generating a nonlinear semi-group endowed with a global Liapunov fonctional. For these systems, the existence of a *maximal attractor*, equal to the unstable set emanating from stationary solutions is established. Finally it is shown that in the case of semilinear parabolic problems, the maximal attractor is a compact set whose Hausdorff dimension is generally finite.

Résumé : On considère divers systèmes d'équations aux dérivées partielles engendrant un semi-groupe non linéaire qui possède une fonction de Liapunov globale. Pour ces systèmes, on établit l'existence d'un *attracteur maximal* qui coïncide avec l'ensemble instable engendré par les solutions stationnaires. On montre enfin que dans le cas des ëquations paraboliques semi-linéaires, l'attracteur maximal est un compact dont la dimension de Hausdorff est en général finie.

C. BARDOS, P. DEGOND. EXISTENCE GLOBALE DES SOLUTIONS DES EQUATIONS DE VLASOV-POISSON.

Abstract : This paper is devoted to the study of the solution of the Vlasov-Poisson equation. A simple model of these equations is given by

$$\begin{cases} \frac{\partial f}{\partial t} + v.\nabla_x f + \nabla\Phi.\nabla_v f = 0 \\ -\Delta\Phi = \int f(x,v,t)dv \end{cases}$$

f is a scalar function of the variables (x,v,t).

We are concern with the existence, regularity and uniqueness of the solution of the Cauchy problem. We emphasize the similarities with the classical wave equation

$$\square u + g|u|^2 u = 0$$

In any dimension there is a weak solution ; when the dimension is small a Sobolev type argument gives the regularity for any smooth initial data, however, when the dimension is large a regularity result can be obtained for the Cauchy problem if the initial datas are small enough. The main ingredient of the proof is the dispersive property of the linear wave equation. For the Vlasov-Poisson equation one can prove (Arseneev [1]) the existence of a weak solution in any space dimension. It is possible for $n \leq 2$ to make use of a Sobolev type argument to obtain the existence and uniqueness of a smooth solution (Ukai-Okabe [9]). However we show that for $n \geq 3$ one can use the local decay of the charge to obtain, for small initial data, a global smooth solution. These results have been announced in a Note of the C.R.A.S (Bardos-Degond [1] and an english written detailed version will appear [2]). Finally, we show that the recent result of Ilner and Shinbrot [7] concerning the Boltzmann equation with initial data close to the vacuum, is also a consequence of the dispersive effect of the linear transport equation. This implies in particular that the regime considered by Ilner and Shinbrot is completely different of the classical situation (small perturbation of a constant Maxwellian) usually studied.

Résumé : On étudie les solutions de l'équation de Vlasov-Poisson. Un modèle simple de ces équations est le suivant

$$\begin{cases} \dfrac{\partial f}{\partial x} + v\nabla_x f + \nabla\Phi.\nabla_v f = 0 \\ -\Delta\Phi = \displaystyle\int f(x,v,t)dv \end{cases}$$

où f est une fonction scalaire des variables (x,v,t). On s'occupe de l'existence, de la régularité et de l'unicité de la solution du problème de Cauchy. On met en valeur les similitudes avec l'équation classique des ondes

$$\square u + g|u|^2 u = 0$$

Quelque soit la dimension, il existe une solution faible ; quand la dimension est petite, un argument de type Sobolev donne la régularité pour une donnée initiale régulière. On obtient un résultat de régularité même quand la dimension est grande, pour le problème de Cauchy a donnée initiale suffisamment petite. L'ingrédient principal de la démonstration est la propriété de dispersion de l'équation linéaire des ondes. Pour l'équation de Vlasov-Poisson, on peut démontrer (Arsennev [1]) l'existence d'une solution faible quelque soit la dimension de l'espace.

Pour $n \leq 2$ on peut utiliser un argument de type Sobolev pour obtenir l'existence et l'unicité d'une solution régulière (Ukai-Okabe [9]). Neanmoins, nous montrons que pour $n \geq 3$ on peut utiliser la décroissance locale de la charge pour obtenir pour données initiales petite, une solution globale régulière ; ces résultats ont été annoncés dans une note au C.R.A.S. (Bardos-Degond [1] et une version anglaise détaillée paraîtra dans [2]). Enfin, nous montrons qu'un résultat récent d'Ilner et Shinbrot [7], concernant l'équation de Boltzmann avec une donnée initiale voisine du vacuum, est aussi une conséquence de l'effet dispersif de l'équation de transport linéaire. Ceci implique en particulier que le regime considéré par Ilner-Shinbrot [7] est tout à fait différent de la situation classique (perturbation petite d'un maxwellian constant) étudié usuellement.

TH. CAZENAVE, A. HARAUX. ON THE NATURE OF FREE OSCILLATIONS ASSOCIATED WITH SOME SEMI-LINEAR WAVE EQUATIONS.

Abstract : The main object of this work is to study the oscillations of the solutions to some semi-linear wave equations of conservative type with homogeneous boundary conditions in a bounded domain Ω of $\mathbb{R}^n$, $n \geq 1$.

If $n=1$ and $\Omega =]0,1[$, it is shown that for g non-decreasing and odd with respect to u, any solution $u \not\equiv 0$ of

$$u \in C(\mathbb{R},H_0^1(\Omega)) \cap C^1(\mathbb{R},L^2(\Omega)), \quad \Box u + g(t,u) = 0$$

is such that for any x_0 outside a finite subset of $\bar{\Omega}$ of the form $\{0, \frac{1}{k},\ldots\frac{m}{k},\ldots 1\}$, the function $u(t,x_0)$ takes both positive and negative values on any interval J such that $|J| > 2$.

If $n > 1$ or if other boundary conditions are taken, some weaker, but quite suggestive oscillation properties are shownin a few typical cases.

Résumé : Le but principal de ce travail est d'étudier les oscillations des solutions de certaines équations des ondes semi-linéaires de type conservatif avec des conditions aux limites homogènes dans un domaine borné Ω de $\mathbb{R}^n$, $n \geq 1$.

Si $n=1$ et $\Omega =]0,1[$, on établit que pour g croissante au sens large et impaire par rapport à u, toute solution $u \neq 0$ de

$u \in C(\mathbb{R},H_0^1(\Omega)) \cap C^1(\mathbb{R},L^2(\Omega))$, $\Box u + g(t,u) = 0$

est telle que pour x_0 quelconque en-dehors d'un sous-ensemble fini de $\bar{\Omega}$ de la forme $\{0, \frac{1}{k},\dots,\frac{m}{k},\dots 1\}$, la fonction $u(t,x_0)$ prend à la fois des valeurs positives et des valeurs négatives sur tout intervalle J tel que $|J| > 2$.

Si $n > 1$ ou si on prend d'autres conditions aux limites, des résultats plus faibles mais très suggestifs d'oscillation sont établis dans quelques cas typiques.

D. de FIGUEIREDO. ON THE UNIQUENESS OF POSITIVE SOLUTIONS OF THE DIRICHLET PROBLEM $-\Delta u = \lambda \sin u$

Abstract : It is proved that for domain Ω with some geometrical properties, the positive solution of the Dirichlet problem $-\Delta u = f(u)$ in Ω, $u = 0$ in $\partial\Omega$, is unique when f has some adequate oscillatory property.

Résumé : On prouve que le problème de Dirichlet pour l'équation $-\Delta u = f(u)$ sur Ω a une seule solution positive pour le cas d'un domaine satisfaisant certaines conditions géométriques et d'une fonction f avec une propriété d'oscillation, comme dans le cas de $f(u) = \lambda \sin u$.

L. FRANK , F. WENDT. ELLIPTIC AND PARABOLIC SINGULAR PERTURBATIONS IN THE KINETIC THEORY OF ENZYMES.

Abstract : A family of semilinear parabolic operator $\mathcal{T}_\varepsilon$ and corresponding elliptic operators $\mathcal{T}_{\varepsilon,\infty}$ is considered, ε being a small parameter and the reduced problem being characterized by the presence of a free boundary. This kind of operators appears in the Kinetic theory of artificial membranes with enzymotic activity.

Résumé : On considère une famille d'opérateurs paraboliques semi-linéaires $\mathcal{T}_\varepsilon$ et celle d'opérateurs elliptiques correspondant $\mathcal{T}_{\varepsilon,\infty}$, ε étant un petit paramètre et le problème réduit étant celui à frontière libre. Ce genre d'opérateurs apparaît en théorie cinétique des membranes artificielles à activité enzymique.

M. GARONNI. ON BILATERAL EVOLUTION PROBLEMS OF NON-VARIATIONAL TYPE.

Abstract : We present some results about bilateral problems related to a second order linear or non-linear elliptic or parabolic operator with the principal part not in divergence form.

We only give the essential information about existence, uniqueness and regularity results and we refer to [7] and [8] for the proofs, the details and for the stochastic characterization.

In this lecture we especially dwell upon the following approximation results : we consider the "regularizing" (variational) problems and we estimate in the uniform norm the order of convergence of the solutions in terms of the order of convergence of the corresponding coefficients.

We sketch the proof in the simplest case, that of strong solutions in the linear case and we refer to [8] and [9] for the proof for generalized solutions in the linear case and for strong solutions in the non linear case, respectively.

Résumé : On présente ici quelques résultats concernant des problèmes bilatéraux qui correspondent à des opérateurs différentiels du deuxième ordre linéaires ou non linéaires elliptiques ou paraboliques dont la partie principale n'est pas en forme de divergence.

On donne seulement les renseignements les plus importants sur l'existence, l'unicité et la régularité et on renvoie à [7] et [8] pour les démonstrations, les détails et la caractérisation stochastique.

Cette conférence concerne surtout les résultats d'approximation ; on considère des problèmes régularisants (variationnels) et l'on évalue, dans la norme uniforme, l'ordre de convergence des solutions en termes de l'ordre de convergence des coefficients (correspondants). On donne une esquisse de la preuve et dans le cas le plus simple, c'est-à-dire dans le cas linéaire et avec des solutions fortes, et on renvoie à [8] et [9] pour le cas linéaire avec des solutions généralisées et pour le cas non linéaire avec des solutions fortes.

A. HARAUX. TWO REMARKS ON HYPERBOLIC DISSIPATIVE PROBLEMS.

Abstract : In the first section we give a new and simple proof of the fundamental result of Amerio-Prouse on boundedness of the energy for the solutions of semi-linear dissipative hyperbolic equations with a forcing term bounded on $\mathbb{R}^+$.

In the second section, we improve a recent result of Babin-Vishik concerning the global behavior of some semi-linear autonomous hyperbolic systems with a linear damping term : more precisely, we show that the trajectories starting from a bounded subset of the energy space are uniformly asymptotic, as $t \to +\infty$, to some compact set consisting in "regular functions".

Résumé : Dans la première partie on donne une nouvelle démonstration simple du résultat fondamental d'Amerio-Prouse sur le caractère borné de l'énergie des solutions d'équations hyperboliques semi-linéaires dissipatives avec second membre borné sur $\mathbb{R}^+$.

Dans la deuxième partie on précise un résultat récent de Babin-Vishik sur le comportement global de certains systèmes hyperboliques semi-linéaires autonomes avec dissipation : plus précisément on montre que les trajectoires issues d'un borné en énergie convergent uniformément lorsque $t \to +\infty$ vers un compact formé de fonctions plus régulières.

V.A. KONDRATIEV, O.A. OLEINIK. ON THE SMOOTHNESS OF WEAK SOLUTIONS OF THE DIRICHLET PROBLEM FOR THE BIHARMONIC EQUATION IN DOMAINS WITH NONREGULAR BOUNDARY.

Abstract : We study the regularity and the behaviour of weak solutions of biharmonic equation with Dirichlet homogeneous and nonhomogeneous boundary conditions. We prove that a weak solution in a closed domain belongs to a Hölder class $C^{1+\delta(\omega)}$ where $\delta(\omega)$ is a root of a transcendental equation with a parameter ω determined by the geometric properties of the domain. It is also shown that the Hölder class can not be improved for domains characterized by the same ω.

Résumé : On étudie la régularité et le comportement des solutions faibles du problème biharmonique avec des conditions au bord de Dirichlet homogènes ou non homogènes. On démontre qu'une solution faible dans un domaine fermé appartient à une classe de Hölder $C^{1+\delta(\omega)}$ où $\delta(\omega)$ est solution d'une certaine équation transcendentale avec un paramètre ω déterminé par les propriétés géométriques du domaine. On montre aussi que cette classe est la meilleure possible pour des domaines caractérisé par le même ω.

P. KREE. NEW BOUNDARY VALUE PROBLEMS CONNECTED WITH MULTIVALUED STOCHASTIC DIFFERENTIAL EQUATIONS.

Abstract : Three results are presented concerning the generalized Fokker-Planck equation (FPE) of the theory of multivalued stochastic equations : see "Mécanique aléatoire" (Dunod-Bordas Paris 1983).

a) Generalized FPE is equivalent with two coupled boundary value problems BVP1 and BVP2.

b) Generalized FPE is interpreted in terms of fluid mechanics. Hence, if the coupling is not trivial, BVP1 is not an usual first kind BVP but a *generalized* first kind BVP.

c) This is proven mathematically. As a corollary, reflected diffusions are constructed such that the free FPE is hypoelliptic and such that the transition probability is not continuous up to the boundary.

Résumé : Trois résultats sont présentés concernant l'équation de Fokker-Planck généralisée (EFPG) de la théorie des équations stochastiques multivoques : voir "Mécanique aléatoire" (Dunod-Bordas, Paris 1983).

a) EFPG est équivalente à deux problèmes aux limites couplés : BVP1 et BVP2.

b) EFPG est interprétée en termes de mécanique de fluids. Cette interprétation "montre" que dans le cas nontriviaux BVP1 n'est pas un problème aux limites usuel de première espèce, mais un problème généralisé de première espèce.

c) Ceci est démontré mathématiquement et donne comme corollaire l'existence de diffusions réfléchies telle que l'EFP de la diffusion libre soit hypoelliptique sur tout l'espace et telles que les probabilités de transition ne soient pas continues jusqu'au bord du domaine.

E.H. LIEB. A LOWER BOUND ON THE CHANDRASEKHAR MASS FOR STELLAR COLLAPSE.

Abstract : The "relativistic" Schrödinger operator H, for N particles interacting via gravitational attraction is investigated. These is always a collapse (i.e. inf spec(H) = $-\infty$) if the coupling constant K, is large enough. It is shown that the critical constant K_c, satisfies (for large N) : $K_c \sim N^{-1}$ for bosons (i.e. symmetric function) and $K_c \sim N^{-2/3}$ for fermions (antisymmetric functions)

Résumé : On étudie l'opérateur "relativiste" de Schrödinger H, pour N particules interagissant via l'attraction gravitationnelle. Il existe toujours un collapse (i.e. inf spec(H) = $-\infty$) si la constante de couplage K est suffisamment grande. On montre que la constante critique K_c, satisfait (pour N grand) : $K_c \sim N^{-1}$ pour les "bossons" (fonctions symétriques) et $K_c \sim N^{-2/3}$ pour les fermions (fonctions antisymétriques).

Y. MEYER. ETUDE D'UN MODELE MATHEMATIQUE ISSU DU CONTROLE DES STRUCTURES SPATIALES DEFORMABLES.

Abstract : Let Ω be a smooth bounded open set in $\mathbb{R}^3$. We study, for $b \in \Omega$ and v(t) given the system

$$\frac{\partial^2 y}{\partial t^2} - \Delta y = v(t)\delta(x-b) \text{ in } \Omega \times]0,T[$$

$$y(x,0) = \frac{\partial y}{\partial t}(x,0) = 0 \text{ in } \Omega$$

$$y = 0 \text{ on } \partial\Omega \times]0,T[$$

We prove that for v in $L^2(0,T)$ the function $x \to y(x,T;v)$ is still in $L^2(\Omega)$ and we precise the functional space of solutions.

Résumé : Soit Ω un ouvert borné régulier dans $\mathbb{R}^3$. On étudie, pour $b \in \Omega$ et v(t) donnés, le système

$$\frac{\partial^2 y}{\partial t^2} - \Delta y = v(t)\delta(x-b) \text{ dans } \Omega \in]0,T[$$

$$y(x,0) = \frac{\partial y}{\partial t}(x,0) = 0 \text{ dans } \Omega$$

$$y = 0 \text{ sur } \partial\Omega \in]0,T[$$

On montre que pour v donné dans $L^2(0,T)$ la fonction $x \to y(x,T;v)$ est dans $L^2(\Omega)$ et on précise l'espace fonctionnel des solutions.

O. PIRONNEAU. TRANSPORT DE MICROSTRUCTURES.

Abstract : On studies Navier-Stokes equations with very oscillating initial data. An asymptotic expansion of homogenization type unables us to find for the averaging field a new system of equations analogous to those of the k-ε turbulence models.

Résumé : On étudie les équations de Navier-Stokes lorsque les conditions initiales sont très oscillantes. Un développement asymptotique du type de ceux qui sont utilisés en homogénéisation permet de trouver pour le champ moyen un nouveau système d'équations semblable à ceux des modèles de turbulence k-ε.

C.H. TAUBES. A BRIEF SURVEY OF THE YANG-MILLS-HIGGS EQUATIONS ON $\mathbb{R}^3$.

Abstract : The author presents an introduction to Yang-Mills-Higgs equations on $\mathbb{R}^3$ and to the physical origin of these problems. A brief survey of known and published results on the question is presented in the end.

Résumé : L'auteur présente une introduction aux problèmes de Yang-Mills-Higgs dans l'espace à trois dimensions ainsi que l'origine physique de ces questions. Un certain nombre de résultats connus sur le sujet sont passés en revue.

R. TEMAM. ATTRACTORS FOR NAVIER-STOKES EQUATIONS.

Abstract : This paper presents the latest results concerning the attractors for the Navier-Stokes equations in space dimension 2 and 3. In the two dimensional case, the existence of a maximal attractor is shown and a bound of its fractal dimension is described, this bound being perhaps optimal. In dimension three, under the assumption that the flow remains smooth in the sense of Leray, it is shown that the dimension of the attractor is that predicted by Kolmogorov theory of turbulence.

Résumé : Cet exposé fait le point sur les attracteurs dans les équations de Navier-Stokes en dimension d'espace deux et trois. En dimension deux, le résultat principal qui est décrit est l'existence d'un attracteur maximal dont la dimension fractale est estimée de manière vraisemblablement optimale en terme d'un nombre sans dimension qui décrit l'écoulement. En dimension trois, sous l'hypothèse que l'écoulement reste régulier au sens de Leray, on montre que la dimension de l'attracteur est celle prédite dans la théorie de Kolmogorov de la turbulence.

A V BABIN & M I VISHIK

Attracteurs maximaux dans les équations aux dérivées partielles

INTRODUCTION. Le texte qui va suivre est le compte-rendu d'une série de conférences données aux Collège de France par le Professeur M.I. Vishik concernant ses travaux en collaboration avec Monsieur A.V. Babin. Le point de vue adopté est légèrement différent de celui des travaux déjà publiés sur ce sujet [1] et [2]. En particulier, nous nous concentrerons essentiellement sur le cas des semi-groupes qui possèdent une fonction de Liapunov globale.

Dans la suite, on considèrera divers systèmes d'équations aux dérivées partielles traduits par une équation d'évolution autonome qui a la propriété de dissiper une certaine fonctionnelle définie sur une partie dense de l'espace des phases.

On cherchera à établir que de tels systèmes possèdent en fait une propriété de dissipation uniforme qui se traduit par la présence d'un *attracteur compact* : c'est-à-dire que toutes les trajectoires définies pour $t \geq 0$ et issues d'un borné convergent *uniformément*, lorsque $t \to +\infty$, vers un certain compact invariant sous l'action du système dynamique associé.

Le problème de déterminer l'attracteur d'un système est différent de celui du comportement à l'infini des trajectoires considérées individuellement dès que ce système possède des *points d'équilibre instables*. En règle générale, la théorie qui va suivre est adaptée à l'étude des systèmes qui, tout en dissipant une certaine fonctionnelle, ne peuvent pas être décrits par un semi-groupe de contractions dans un espace de Banach.

La compacité de l'attracteur, parfois délicate à établir, est liée heuristiquement au caractère irréversible du système. En fait, dans de nombreux cas intéressants en pratique, on arrive même à établir que la dimension de Hausdorff de l'attracteur est finie. (cf. [2] pour les démonstrations complètes). Signalons également que dans le cas générique" d'un système possédant un nombre fini de points d'équilibre, l'attracteur peut être décrit de manière beaucoup plus complète (cf. [1] et pour les détails). Enfin, pour un exposé détaillé des résultats relatifs à l'équation de Navier-Stokes, cf. [13].

1. GENERALITES - ATTRACTEURS COMPACTS.

Dans ce paragraphe, on désigne par E un espace de Banach réel et S_t un semi-groupe d'opérateurs *continus* : $E \to E$ tels que pour tout $u_o \in E$ la fonction $t \to S_t u_o$ soit continue.

Définition 1.1. Soit $\mathcal{A} \subset E$ une partie *fermée bornée*. On dit que $\mathcal{A}$ est un *attracteur maximal* pour S_t dans E si

1) $\forall t \geq 0, \quad S_t \mathcal{A} = \mathcal{A}$

2) $\forall B \subset E$, B *borné* on a

$$\sup_{x \in B} \{\inf_{y \in \mathcal{A}} \|S_t x - y\|\} \to 0 \quad \text{lorsque} \quad t \to +\infty$$

Proposition 1.2. Si $\{S_t\}$ possède un attracteur maximal, celui-ci est unique.

Démonstration. Il résulte clairement de la définition que pour tout fermé borné $\mathcal{B} \subset E$ tel que $\forall t \geq 0$, $S_t \mathcal{B} = \mathcal{B}$, on a

$$\sup_{x \in \mathcal{B}} d(x, \mathcal{A}) = 0 \Rightarrow \mathcal{B} \subset \mathcal{A} .$$

Définition 1.3. Soit B_o un fermé borné de E. On dit que B_o est *absorbant* si $\forall\, B \subset E$, B borné, $\exists\, T(B)$ tel que

$$\forall t \geq T(B), \quad S_t(B) \subset B_o.$$

Théorème 1.4. Si $\{S_t\}$ possède un ensemble *compact* absorbant, alors il possède également un *attracteur maximal compact*.

Démonstration. Pour tout $\tau > 0$ on pose

$$F_\tau = \overline{\bigcup_{t \geq \tau} S_t(B_o)}.$$

Puisque $S_t(B_o) \subset B_o$ pour t assez grand, il est clair que F_τ est *compact*. Soit

$$\mathcal{A} = \bigcap_{\tau>0} F_\tau .$$

Alors $\mathcal{A}$ est l'attracteur maximal cherché.

En effet

1) Pour tout $\theta > 0$, on a par continuité de S_θ

$$S_\theta(F_\tau) \subset S_\theta(\overline{\bigcup_{t\geq\tau} S_t(B_o)}) \subset F_{\theta+\tau}$$

Comme F_τ décroît avec τ, il en résulte que

$$S_\theta(\mathcal{A}) \subset \mathcal{A} .$$

D'autre part si $x \in \mathcal{A}$ on peut supposer

$$x = \lim_E (S_{t_j} x_j),\ x_j \in B_o,\ t_j \to +\infty .$$

Soit $t'_j = t_j - \theta$. Alors

$$x = \lim_E S_\theta[S_{t'_j} x_j]$$

Comme $S_{t'_j} x_j \in B_o$ pour j assez grand, on peut quitte à extraire une sous-suite supposer que

$$S_{t'_j} x_j \to z \text{ dans } E \text{ lorsque } j \to +\infty .$$

Alors $z \in \mathcal{A}$ et $S_\theta z = x$.

2) Supposons, par l'absurde, que l'on ait

$$\limsup_{t\to+\infty} \sup_{x\in B} \{\inf_{y\in\mathcal{A}} \|S_t x - y\|\} = 2\varepsilon > 0 .$$

On peut alors supposer, avec $t_j \to +\infty$, $t_{j+1} > t_j$ et $x_j \in B$, que

$$d(S_{t_j} x_j, \mathcal{A}) \geq \varepsilon, \quad \forall j \in \mathbb{N} .$$

Pour $j \geq j_o$, on a $S_{t_j} x_j = S_{t_j - t_{j_o}} (S_{t_{j_o}} x_j)$ et $S_{t_{j_o}} x_j \in B_o$, d'où quitte à extraire une sous-suite de $\{t_j\}$:

$$\lim_E S_{t_j} x_j = x \in \mathcal{A}$$

Cette contradiction achève la démonstration du théorème 1.4. □

Remarque. Le théorème 1.4 s'applique en particulier à tout semi-groupe $\{S_t\}$ *compact* ayant un borné absorbant, ce qui est fréquent pour les systèmes paraboliques (cf. exemples de [2]).

Soit maintenant Y une partie *fermée* de E. La définition suivante nous sera constamment utile.

Définition 1.5. On appelle ensemble instable engendré par Y dans E (sous l'action de S_t) le sous-ensemble de E défini par $M_+(Y) = \{z \in E,\ u \in C(\mathbb{R},E)$ avec $u(0) = z$, $S_t u(\tau) = u(\tau+t)$, $\forall (t,\tau) \in \mathbb{R}^+ \times \mathbb{R}$ et $\text{dist}_E(u(\tau),Y) \to 0$ lorsque $\tau \to -\infty\}$.

Proposition 1.6. Si $M_+(\{y\}) \neq \emptyset$, on a

$$S_t y = y, \quad \forall t \geq 0.$$

Démonstration. Soit $z \in M_+(\{y\})$. On a donc $\lim_{\tau \to -\infty} \|u(\tau) - y\| = 0$ et $u(\tau+t) = S_t u(\tau)$. Donc $S_t y = \lim_{\tau \to -\infty} u(\tau+t) = \lim_{\tau \to -\infty} u(\tau) = y$.

Remarque 1.7. Considérons le cas particulier où $E = \mathbb{R}$ et $\{S_t\}$ est le semi-groupe dans $\mathbb{R}$ engendré par l'équation différentielle ordinaire

$$u' = u - u^3.$$

Le système possède 3 points d'équilibre 0, -1 et 1. Toute trajectoire de l'équation de donnée initiale $u_o \neq 0$ converge vers l'un des deux points -1 et 1, tandis que 0 est un point d'équilibre instable.

Dans ce cas particulier très simple, *l'attracteur* $\mathcal{A}$ est donné par $\mathcal{A} = [-1,+1]$. On remarque que $]-1,+1[= M_+(\{0\})$. Cet exemple permettra de comprendre plus facilement ces résultats généraux des paragraphes 2 et 3.

Proposition 1.8. Si $Y = \{y_1,\ldots, y_n\}$ on a

$$M_+(Y) = \bigcup_{i=1}^{n} M_+(\{y_i\}).$$

Démonstration. Soit $z \in M_+(Y)$ et u la trajectoire associée. Pour tout $\tau_o \leq 0$, $\Gamma_{\tau_o} = \overline{\bigcup_{\tau<\tau_o} u(\tau)}$ est un compact connexe. Donc $\gamma = \bigcap_{\sigma<0} \Gamma_\sigma$ est encore compact et connexe.

Comme $\gamma \subset \{y_1,\ldots, y_n\}$, on a en fait $\gamma = \{y_j\}$ pour un certain $j \in \{1,\ldots,n\}$.

Remarque 1.9. La définition de $M_+(Y)$ fait intervenir des "trajectoires" de S_t définies sur $]-\infty,+\infty[$. Ceci nous amène à une nouvelle définition.

Définition 1.10. On appelle *trajectoire complète* de $\{S_t\}$ une courbe $u \in C(\mathbb{R},E)$ telle que l'on ait

$$\forall t \geq 0, \quad \tau \in \mathbb{R}, \quad u(\tau+t) = S_t u(\tau).$$

Proposition 1.11. Soit $X \subset E$ tel que

$$\forall t \geq 0, \quad S_t X = X \quad \text{(invariance)}$$

Alors $\forall x \in X$, il existe une trajectoire complète u de $\{S_t\}$ telle que $u(0) = x$ et $u(\tau) \in X$, $\quad \forall \tau \in \mathbb{R}$.

Démonstration. Définissons par récurrence une suite $\{x_n\}_{n\geq 0}$ telle que $S_1(x_{n+1}) = x_n$ et $x_0 = x$. Alors on peut poser $u_n(\tau) = S_{n+\tau} x_n$ pour $\tau \geq -n$. On vérifie que si $m > n$, alors $u_m = u_n$ sur $[-n,+\infty[$. Finalement la fonction définie par $u(\tau) = u_n(\tau)$ si $\tau \geq -n$ répond aux conditions exigées.

Remarque 1.12. Si $\{S_t\}$ possède un attracteur $\mathcal{A}$ alors toute trajectoire complète *bornée* $u(\tau)$ est telle que $\forall\tau \in \mathbb{R}$, $u(\tau) \in \mathcal{A}$.

En effet, on a $u(\tau) = S_n u(\tau-n)$ et $u(\tau-n)$ reste dans un borné B lorsque n décrit $\mathbb{N}$.

2. SEMI-GROUPES COMPACTS MUNIS D'UNE FONCTION DE LIAPUNOV GLOBALE.

Soit X une partie de E telle que $\forall t \geq 0$, $S_t X \subset X$.

Définition 2.1. Une fonction $\phi : X \to \mathbb{R}$ est dite *fonction de Liapunov* de $\{S_t\}$ sur X si l'on a les propriétés suivantes

1) $\forall u \in X$, la fonction $t \to \phi(S_t u)$ est décroissante au sens large.

2) Si $z \in X$ est tel que $\phi(S_t z) \equiv \phi(z)$ alors $S_t z = z$, $\forall t \geq 0$.

Dans la suite, on notera $\mathcal{M}(X)$ l'ensemble

$$\{z \in X, \forall t \geq 0, \ S_t z = z\}.$$

Théorème 2.2. *Soit* $X \subset E$ *tel que*

1) X *compact et* $S_t X = X$, $\forall t \geq 0$

2) *Il existe une fonction de Liapunov* $\phi : X \to \mathbb{R}$ *continue sur* X *pour la topologie de* E.

Alors on a $X \subset M_+(\mathcal{M}(X))$.

De plus, si $\mathcal{M}(X) = \{z_1, z_2, \ldots, z_n\}$ *alors* $X \subset \bigcup_{i=1}^{n} M_+(\{z_i\})$.

Démonstration. Soit $x \in X$ et u une trajectoire complète de $\{S_t\}$, contenue dans X, telle que $u(0) = x$. Posons

$$Z = \bigcap_{\tau_0 < 0} \overline{\bigcup_{\tau < \tau_0} u(\tau)}^{E}.$$

Alors Z possède les propriétés suivantes

1. $S_t Z = Z$, $\forall t \geq 0$.

2. La fonction ϕ est constante sur Z, plus précisément

$$\forall z \in Z, \quad \phi(z) = \lim_{\tau \to -\infty} \phi(u(\tau)).$$

En particulier, on a $Z \subset \mathcal{M}(X)$.

Or il est clair que $d(u(\tau),Z) \to 0$ lorsque $\tau \to -\infty$. On a donc

$$x \in M_+(Z) \subset M_+(\mathcal{M}(X)).$$

Ceci achève la démonstration du théorème 2.2.

Remarque 2.3. On n'a pas en général l'égalité $X = M_+(\mathcal{M}(X))$ comme le montre le cas particulier $X = \{y\}$ avec y un point d'équilibre *instable*. Cependant, le théorème 2.2 permet d'établir le résultat fondamental suivant.

Corollaire 2.4. *Soit* $\{S_t\}$ *un semi-groupe possèdant un attracteur maximal compact* $\mathcal{A}$ *. Si* $\{S_t\}$ *a une fonction de Liapunov continue sur* $\mathcal{A}$ *pour la topologie de* E *, alors*

$$\mathcal{A} = M_+(\mathcal{M}(E)).$$

Démonstration. Le théorème 2.2 donne

$$\mathcal{A} \subset M_+(\mathcal{M}(\mathcal{A})) \subset M_+(\mathcal{M}(E)).$$

Grâce à la remarque 1.12, on a d'autre part $M_+(\mathcal{M}(E)) \subset \mathcal{A}$.

Remarque 2.5. Si S_t est *compact* pour tout $t > 0$ et si $\{S_t\}$ possède un borné absorbant B_o, alors en fait $S_1(B_o)$ est un compact absorbant. D'après le théorème 1.4, $\{S_t\}$ possède alors un *attracteur maximal compact*. Le corollaire 2.4 est applicable à divers systèmes paraboliques non linéaires dans un ouvert Ω borné de $\mathbb{R}^n$.

Exemple 1. On considère le problème semi-linéaire

$$\begin{cases} \partial_t u = \Delta u - f(u) - g(x) \\ u|_{\partial\Omega} = 0 \end{cases}$$

Si $f \in C^1(\mathbb{R})$ avec $f' \geq - C$ sur $\mathbb{R}$, cette équation engendre un semi-groupe $\{S_t\}$ d'opérateurs continus (en fait, lipschitziens) sur $E = L^2(\Omega)$ dès que $g \in L^2(\Omega)$.

Posons $F(u) = \int_0^u f(s)ds$ et définissons

$$\phi(u) = \int_\Omega \{\tfrac{1}{2}|\nabla u|^2 + F(u) + g(x)u\}dx$$

Il est clair que pour tout $M > 0$, la fonction ϕ est bornée et continue sur

$$W_M = \{u \in H^2(\Omega),\ f(u) \in L^2(\Omega),\ \|u\|_{H^2(\Omega)} + \|f(u)\|_{L^2(\Omega)} \leq M\}$$

lorsqu'on munit W_M de la topologie de $L^2(\Omega)$.

D'autre part pour toute solution régulière $u(t,x)$ on a

$$\frac{d}{dt}(\phi(u(t))) = - \int_\Omega |\partial_t u|^2 dx = - \int_\Omega |\Delta u - f(u) - g(x)|^2 dx.$$

En particulier, $\phi(u(t))$ est une fonction de Liapunov sur tout ensemble $X \subset W_M$ tel que $S_t X \subset X$.

Si on suppose que l'on a de plus $f(u)u \geq (-\lambda_1 + \varepsilon)|u|^2 - C$, il est immédiat de vérifier que $\{S_t\}$ possède un borné absorbant dans E.

On en déduit par des manipulations classiques que W_M est absorbant pour $M > 0$ assez grand. Il en résulte que $\{S_t\}$ a un attracteur compact qui est en fait contenu dans $W_{M'}$, pour M' assez grand. (Remarquer pour cela que si u est une solution de l'équation, $|\partial_t u|_2$ augmente de façon contrôlée par une exponentielle et utiliser $|\partial_t u|_2^2 = - \frac{d}{dt}(\phi(u(t)))$).

Donc ϕ est une fonction de Liapunov qui est continue sur $\mathcal{A}$ muni de la topologie de $E = L^2(\Omega)$.

Exemple 2. On prend $\Omega =]0,2\pi[^n = T^n$ muni des conditions aux limites périodiques et on considère l'équation

$$\partial_t u = \sum_{i=1}^n \partial_i [a_i(\nabla u)] - f(u) + \lambda u - g(x)$$

où $a_i(\xi) = \partial_i a(\zeta)$, a étant une fonction convexe de classe C^2 sur $\mathbb{R}^n$ telle que

$$\mu_0|\xi|^2 \le \sum_{i,j} a_{ij}(\zeta)\xi_i\xi_j \le \mu_1(|\xi|^2)$$

$$\forall\ (\zeta,\xi) \in \mathbb{R}^n \times \mathbb{R}^n$$

(avec $0 < \mu_0 \le \mu_1 < +\infty$)

$f \in C^1(\mathbb{R})$ vérifie $f'(u) \ge c|u|^\alpha - C$

$c > 0, \quad \alpha > 0.$

On prend $E = L^2(T^n)$ et

$$\phi(u) = \int_\Omega \{a(\nabla u) + F(u) - \frac{\lambda^2}{2} u^2 + g(x)u(x)\}dx$$

On démontre comme dans l'exemple 1 l'existence d'un attracteur maximal $\mathcal{A} \subset H^2(T^n)$.

De plus ϕ est continue sur $\mathcal{A}$ pour la topologie de E.

Exemple 3. On peut étudier de la même façon l'équation

$$\begin{cases} \partial_t u = \phi(\Delta u - f(u) - g(x)) \\ u|_{\partial\Omega} = 0 \end{cases}$$

où Ω est un ouvert borné de $\mathbb{R}^n$, $n \le 3$ et $\phi \in C^4 \cap W^{2,\infty}(\mathbb{R})$ vérifie $\phi(0) = 0$ et

$$0 < \mu_0 \le \phi'(z) \le \mu_1, \quad \forall z \in \mathbb{R}$$

tandis que $f \in C^4(\mathbb{R})$, $g \in C^{1,\alpha}(\Omega), \alpha > 0$.

Dans ce cas on doit travailler dans

$$E = \{v \in C^{2,\alpha}(\Omega), \quad v|_{\partial\Omega} = \Delta v|_{\partial\Omega} = 0$$

après avoir posé $v = u - h$, h étant la solution de : $\Delta h = g$, $h|_{\partial\Omega} = 0$.

La fonctionnelle

$$\phi(u) = \int_\Omega \{|\nabla u|^2 + F(u) + g(x)u(x)\}\, dx$$

est de Liapunov pour le système avant transformation.

Pour les détails, se reporter à [1], p. 458-461.

Remarques 2.6.

a) Le théorème 1.4 permet de démontrer l'existence d'un attracteur maximal compact même lorsque le corollaire 2.4 n'est pas applicable. Ceci peut se produire pour diverses raisons : soit qu'aucune fonction de Liapunov globale n'apparaisse naturellement (cas des *systèmes*, ou exemples 1 et 2 perturbés par l'addition d'un opérateur du premier ordre), soit que $\{S_t\}$ ne soit pas différentiable (cas de l'exemple 2 où on relaxe les conditions sur $a(\zeta)$, contre-exemple à la différentiabilité dû a Kouksine). Dans le deuxième cas on ne sait plus établir la continuité de ϕ sur $\mathcal{A}$ muni de la topologie de E.

b) D'autres exemples où le théorème 1.4 serait applicable sont traités dans [1]. Nous omettons volontairement ceux où E est du type $C^{2,\alpha}$ en raison des complications techniques.

3. (E_1,E)-ATTRACTEURS ET APPLICATIONS.

Dans ce paragraphe, E et E_1 désignent deux espaces de Banach réel réflexifs tels que

$$E_1 \hookrightarrow E \qquad \text{(injection } compacte\text{)}$$

On considère un semi-groupe d'opérateurs continus $S_t : E \to E$ comme au §2. On suppose de plus que l'on a

$$\forall t \geq 0, \quad S_t(E_1) \subset E_1$$

et que $\{S_t\}$ est *uniformément borné sur* E_1, c'est-à-dire que si l'on désigne par B_R^1 la boule de centre 0 et de rayon R dans E_1, on a

$$\forall R > 0, \quad \exists K(R),$$

$$\forall t \geq 0, \quad S_t(B_R^1) \subset B_{K(R)}^1$$

Proposition 3.1. Sous les hypothèses précédentes, pour toute partie bornée V de E_1, il existe un borné X(V) de E_1 tel que

1) $V \subset X(V)$

2) X(V) fermé dans E

3) $S_t X(V) \subset X(V)$, $\forall t \geq 0$.

Démonstration. Il suffit de prendre

$$X(V) = \overline{\bigcup_{t\geq 0} S_t V}^E$$

Théorème 3.2. *Pour tout* $R > 0$, *l'ensemble* $\mathcal{A}_R = \bigcap_{\tau>0} \overline{\bigcup_{t\geq\tau} S_t(X(B_R^1))}^E$ *est borné dans* E_1 *et vérifie*

1) $S_t(\mathcal{A}_R) = \mathcal{A}_R, \quad \forall t \geq 0$

2) $\forall B \subset B_R^1, \quad \lim_{t\to+\infty} \operatorname{Sup}_{x\in B} \{\operatorname{Inf}_{y\in \mathcal{A}_R} \|S_t x-y\|_E\} = 0.$

De plus, si $\mathcal{A}^\circ = \bigcup_{R>0} \mathcal{A}_R$ *est borné dans* E_1, *il est alors fermé dans* E *et l'on a*

1) $S_t(\mathcal{A}^\circ) = \mathcal{A}^\circ, \quad \forall t \geq 0$

2) $\forall B$ *borné dans* E_1,

$$\lim_{t\to+\infty} \operatorname{Sup}_{x\in B} \{\operatorname{Inf}_{y\in\mathcal{A}^\circ} \|S_t x-y\|_E\} = 0.$$

Démonstration. La première partie du théorème 3.2 se démontre de façon analogue au théorème 1.4. Pour la deuxième partie, il suffit de remarquer que si $\mathcal{A}^\circ \subset B_M^1$ avec M fini on a en fait $\mathcal{A}^\circ \subset \mathcal{A}_M$ et donc $\mathcal{A}^\circ = \mathcal{A}_M$.

Remarque : On dira alors que $\mathcal{A}°$ est un (E_1,E)-attracteur pour $\{S_t\}$.

Définition 3.3. Soit $Y \subset E_1$. On pose

$M^1_+(Y) = \{z \in E,\ \exists\, u \in C(\mathbb{R},E) \cap L^\infty(\mathbb{R},E_1)$ avec $u(0) = z,\ S_t u(\tau) = u(\tau+t)$, $\forall (t,\tau) \in \mathbb{R}^+ \times \mathbb{R}$ et $\text{dist}_E(u(\tau),Y) \to 0$ lorsque $\tau \to -\infty\}$.

Théorème 3.4. *Soit* $\phi(u)$ *une fonction de Liapunov (au sens de la définition* 2.1*) de* $\{S_t\}$ *sur* $\mathcal{A}°$ *et qui est continue sur chaque* $\mathcal{A}_R$ *pour la topologie de* E. *On a alors*

$$\mathcal{A}° = M^1_+(\mathcal{M}(E_1))$$

Corollaire 3.5. *Si* $\{S_t\}$ *restreint à* E_1 *possède un ensemble absorbant borné au sens de* E_1*, l'ensemble* $\mathcal{A}° = M^1_+(\ \mathcal{M}(E_1))$ *est un fermé borné de* E_1 *qui est l'unique fermé borné* $\mathcal{A}$ *de* E_1 *à vérifier les conditions*

1) $S_t(\mathcal{A}) = \mathcal{A}\ ,\quad \forall t \geq 0$

2) $\forall B$ borné dans E_1

$$\lim_{t\to+\infty} \sup_{x\in B} \{\inf_{y\in} \|S_t x-y\|_E\} = 0$$

Les démonstrations du théorème 3.4 et du corollaire 3.5 ne posent pas de problème particulier compte tenu du théorème 3.2 et de la démonstration du théorème 2.2.

Exemple d'application.

Soit Ω un ouvert borné de $\mathbb{R}^n$ $n \leq 3$ et considérons le problème

$$\begin{cases} \partial_t^2 u - \Delta u + \varepsilon \partial_t u + f(u) = g(x) \\ u|_{\partial\Omega} = 0 \end{cases}$$

On suppose $g \in L^2(\Omega)$, $\varepsilon > 0$ et $f \in C^1(\mathbb{R})$ avec

$$f'(u) \geq -K, \qquad \forall u \in \mathbb{R}$$

$$F(u) = \int_0^u f(s)ds \geq -C, \qquad \forall u \in \mathbb{R}.$$

On suppose de plus que f vérifie les conditions :

$$|f'(u)| \leq C(1+|u|^k) \text{ avec } k \geq 0 \text{ si } n \leq 2$$

$$|f'(u)| \leq C(1+|u|^2) \quad \text{si } n = 3.$$

Le résultat suivant est démontré dans [1] sous des conditions légèrement plus restrictives.

Proposition 3.6. Sous les conditions précédentes, l'équation ci-dessus engendre un semi-groupe de transformations continues $S_t : E \to E$ avec

$$E = H_0^1(\Omega) \times L^2(\Omega) \text{ et}$$

$$S_t(u_o,p_o) = (u(t,.), u_t(t,.)) \text{ où } u(t,x)$$

est la solution du problème de Cauchy associé aux données initiales (u_o,p_o). De plus si $E_1 = H^2(\Omega) \cap H_0^1(\Omega) \times H_0^1(\Omega)$, S_t envoie E_1 dans E_1 et est *uniformément borné sur* E_1.

Enfin on a

$$\mathcal{A} = M_+^1(\mathcal{M}(E)) = M_+^1(\mathcal{M}(E_1))$$

au sens du corollaire 3.5.

Démonstration. La première partie de l'énoncé de la proposition 3.6 (semi-groupe S_t dans E) est classique. Le fait que $S_t : E_1 \to E_1$ est uniformément borné est démontré dans [1] lorsque n=3 et $|f'(u)| \leq C(1+|u|^{2-\gamma})$, $\gamma > 0$ où les auteurs émettent la conjecture que cela reste vrai lorsque $\gamma = 0$.

Démontrons ce dernier point en donnant des calculs formels. La fonction $p(t,x) = \partial_t u(t,x)$ vérifie l'équation

$$\partial_t^2 p = -\varepsilon\partial_t p + \Delta p - f'(u)p.$$

Posons $E(t) = \frac{1}{2}|p|_2^2 + \frac{1}{2}|\nabla u|_2^2$

$$E_1(t) = \frac{1}{2}|\partial_t p|_2^2 + \frac{1}{2}|\nabla p|_2^2$$

où $|w|_2$ désigne la norme de w dans $L^2(\Omega)$. Remarquons d'abord que $E_1(t)$ vérifie formellement l'équation différentielle

$$\frac{dE_1}{dt} = \langle\partial_t p,\ \partial_t^2 p - \Delta p\rangle$$

$$= -\varepsilon|\partial_t p|_2^2 - \int_\Omega f'(u)p\partial_t p\,dx$$

On a donc en particulier :

$$\frac{dE_1}{dt} \leq \frac{1}{4\varepsilon}\int_\Omega f'^2(u)p^2dx \leq \frac{1}{4\varepsilon}\ |p(t)|_2^2|f'(u)|_\infty^2\ .$$

Puisqu'on s'est placé dans le cas n=3, on a par hypothèse $|f'(u)|_\infty^2 \leq C(1+|u|_\infty^4)$, et d'après l'inégalité de Gagliardo-Nirenberg en dimension 3

$$|u|_\infty^2 \leq D|u|_{H^2(\Omega)}|u|_{H^1(\Omega)}\ .$$

Comme $|u|_{H^1(\Omega)} \leq M\sqrt{E(t)} \leq P(E(o))$

on trouve

$$\frac{dE_1}{dt} \leq C(E(o))|p(t)|_2^2(1+|u(t)|_{H^2(\Omega)}^2).$$

En reprenant l'équation et puisque f' est minoré on majore $1 + |u(t)|_{H^2(\Omega)}^2$ par $K(E(o))(1+|\partial_t p|_2^2)$.

Or on sait que $\int_0^{+\infty}|p(t)|_2^2dt \leq C'(E(o))$.

Finalement avec $h(t) = |p(t)|_2^2$ on trouve

$$1 + E_1(t) \leq [1+E_1(o)] \exp\{\int_0^{+\infty} CK(E(o))h(t)dt)$$

Il en résulte bien que $\{S_t\}$ est uniformément borné dans E_1.

Dans [1] le fait que $\mathcal{A} = M_+^1(\mathcal{M}(E))$ n'est établi que dans le cas dit "générique" où $\mathcal{M}(E)$ est fini en utilisant le caractère hyperbolique (cf. §4 plus loin) des points d'équilibre. Grâce au corollaire 3.5 nous pouvons établir ce résultat en toute généralité si nous savons démontrer que $\{S_t\}$ a un borné absorbant au sens de E_1.

Or ceci est possible par un argument très voisin du calcul que nous venons d'effectuer. Soit en effet $\eta > 0$ et posons

$$V(t) = \frac{1}{2}|\partial_t p|_2^2 + \frac{1}{2}|\nabla p|_2^2 + \eta(p,\partial_t p)_2$$

On a

$$V'(t) = (-\varepsilon+\eta)|\partial_t p|_2^2 + \eta < p, -\varepsilon\partial_t p + \Delta p - f'(u)p> - \int_\Omega f'(u)p\partial_t p\, dx$$

$$= (-\varepsilon+\eta)|\partial_t p|_2^2 - \varepsilon\eta < p, \partial_t p> - \eta|\nabla p|^2 - \eta \int_\Omega f'u)p^2$$

$$- \int_\Omega f'(u)p\partial_t p\, dx$$

Pour η assez petit on en déduit

$$V'(t) \leq - \delta\, V(t) + C(E(0))|p(t)|_2^2(V(t)+1).$$

Il est immédiat d'en déduire :

$$V(t) \leq M(E(0))e^{-\delta t}V(0) + C'(E(0)).$$

D'autre part l'existence d'un borné absorbant dans E est classique.

Ceci achève la démonstration de la proposition 3.6 lorsque n=3. Les autres cas se traitent d'une façon analogue.

Remarque 3.7. Il serait intéressant de savoir si $\mathcal{A}$ est aussi un attracteur pour les bornés au sens de E. Il semble plausible que cela soit vérifié au moins si n=3, $|f'(u)| \leq C(1+|u|^{2-\gamma})$ avec $\gamma > 0$ et si on étudie la convergence des bornés de E dans la topologie plus faible associée à l'espace $L^2(\Omega) \times H^{-1}(\Omega)$.

Autres exemples. On peut étudier de la même façon au moyen de la notion de (E_1,E) attracteur le comportement global de systèmes tels que

1) $$\begin{cases} \partial_t u = \sum \partial_i(a_i(\nabla u)) - \frac{\partial F}{\partial u}(u,v) - g_1, & x \in T^n \\ \partial_t u = - \frac{\partial F}{\partial v}(u,v) & , \quad x \in T^n \end{cases}$$

Dans ce cas $E = L^2 \times L^2$ et $E_1 = H^1 \times H^1$

2) $$\begin{cases} \partial_t u - \partial_x(a(\partial_x u)) - \lambda\partial_t^2\partial_x^2 u + \varepsilon\, \partial_t u \\ \qquad\qquad - \frac{\partial F}{\partial u}(u,v) = g_1 & x \in]0,1[\\ \partial_t v - \partial_x^2 v \quad - \frac{\partial F}{\partial v}(u,v) = g_2 & x \in]0,1[\\ u|_{x=0} = u|_{x=1=0} \qquad v|_{x=0} = v|_{x=1=0} \end{cases}$$

Dans ce cas les espaces sont compliqués.

(Travaux de Berkaliev, à paraître).

4. ATTRACTEURS REGULIERS - DIMENSION DE HAUSDORFF DE L'ATTRACTEUR.

Nous abordons maintenant les aspects les plus délicats et les plus profonds de cette théorie. La notion "d'attracteur régulier" est expliquée en détail dans [1] avec les démonstrations complètes. (cf. notamment les énoncés des théorèmes 1.1 et 1.2 et les démonstrations des §4 et 5 de [1]). En ce qui concerne la dimension de Hausdorff, les démonstrations détaillées sont écrites dans [2], §3 à 9.

Nous nous contenterons ici d'une revue très succinte des principales idées.

a) Situation générique et attracteurs réguliers.

Soit $S_t : E \to E$ le semi-groupe engendré par une équation d'évolution

$$\begin{cases} \partial_t u = Au \\ u|_{t=0} = u_o \end{cases}$$

Pour étudier la stabilité d'une trajectoire $u(t) = S_t u_o$, on forme "l'équation aux variations"

$$\begin{cases} \partial_t v = A_u'(u(t))v(t) \\ v|_{t=0} = v_o \end{cases}$$

Il est classique que lorsque cette opération a un sens (par exemple si $E = \mathbb{R}^n$ et $A \in C^1(E)$), l'équation aux variations représente l'effet de la dérivée de Fréchet de S_t sur les éléments $v \in E$, en ce sens que

$S_t'(u_o)v_o = v(t)$, solution de l'équation aux variations de donnée initiale v_o.

Lorsque $u(t) \equiv z$ avec $S_t z = z$, $\forall t \geq 0$, l'opérateur $A_u'(u(t))$ se réduit à $A_u'(z)$ et donc $S_t'(z)$ est indépendant de t.

On suppose maintenant que $z \in E$ est un *point stationnaire* de $\{S_t\}$ et que dans un voisinage $\mathcal{O}$ de z, les opérateurs $S_t : E \to E$ sont différentiables avec

$$\forall u_1 \in \mathcal{O}, \; \forall u_2 \in \mathcal{O}, \; \|S_t'(u_1) - S_t'(u_2)\| \leq C\|u_1 - u_2\|^\alpha, \text{ avec } 0 < \alpha \leq 1.$$

Définition. On dit que z est un point stationnaire hyperbolique si cette hypothèse de régularité est satisfaite et

1) $\sigma(S_t'(z)) \cap \{|\zeta| = 1\} = \emptyset$

2) $\sigma(S_t'(z)) \cap \{|\zeta| > 1\}$ est *fini et purement ponctuel.*

Soit E_+ la réunion des sous-espaces propres associés et E_- le sous-espace invariant associé à $\sigma(S_t'(z)) \cap \{|\zeta| < 1\}$.

On a $E = E_- \oplus E_+$, avec $\dim(E_+) < +\infty$. On sait (travaux de Hartman [6], Wells [14], Marsden et Mac-Cracken [9]) qu'il existe au moins localement des variétés différentiables $M_+^o(z)$ et $M_-^o(z)$ tangentes à E_+ et E_- respectivement telles que si u_o est assez voisin de z, on a $\operatorname*{dist}_{t\to+\infty}(u(t), M_+^o(z)) \le Ce^{-\delta t}$, $\delta > 0$.

$M_+^o(z)$ est appelée *variété instable* au voisinage du point stationnaire hyperbolique z. En effet les solutions partant d'un point proche de z situé sur $M_+^o(z)$ ont tendance à s'écarter de z lorsque t augmente. La variété instable est de dimension finie par définition de E_+ et 2). Le résultat suivant, de démonstration délicate, est établi dans [1].

Théorème 4.1. *On a en fait*

$$M_+^o(z) \cap B(z,r) = M_+(\{z\}) \cap B(z,r)$$

pour tout $r > 0$ *assez petit.*

Nous ne donnerons pas la démonstration de ce théorème qui est fondamental dans l'étude des "attracteurs réguliers". Ce résultat permet d'établir le théorème suivant qui évite d'avoir à vérifier l'existence d'un borné absorbant.

Théorème 4.2. *Soit* $S_t : E \to E$ *un semi-groupe compact et uniformément borné sur* E *(i.e.* $S_t(B_R) \subset B_{C(R)}$, $\forall t \ge 0$*) tel que* $(u,t) \to S_t u$ *est continu sur* $E \times \mathbb{R}^+$ *à valeurs dans* E.

On suppose que S_t *possède une fonction de Liapunov et qu'il n'a qu'un nombre fini de points stationnaires tous hyperboliques.*

Soit $\{z_j\}_{j\in\{1,..k\}}$ *l'ensemble des points stationnaires. Alors* S_t *possède un attracteur maximal compact et on a*

$$\mathcal{A} = \bigcup_{1\le j\le k} M^+(\{z_j\}).$$

Dans le cas de l'équation parabolique

$$\begin{cases} \partial_t u = \Delta u - f(u) - g(x) \\ u|_{\partial\Omega} = 0 \end{cases}$$

on sait établir par des arguments classiques (Théorème de Sard-Smale) que pour toutes les fonctions g d'un ouvert dense de $C_0^{0,\alpha}(\bar{\Omega})$ le problème : $u \in H^2 \cap H_0^1(\Omega)$, $\Delta u = f(u) + g(x)$ n'a qu'un nombre fini de solutions lorsque f vérifie des conditions de régularité et de "coercivité à l'infini".

Ainsi, *génériquement*, le système ne possède qu'un nombre fini de points stationnaires. On démontre le caractère *hyperbolique* de ces points stationnaires par une étude détaillée de *l'équation aux variations*.

Ces arguments peuvent également être appliqués à d'autres systèmes. Pour les résultats détaillés et les démonstrations, nous renvoyons à [1].

Remarque 4.3. a) On connaît en fait davantage de précisions sur la structure de $\mathcal{A}$ et les situations relatives des $M^+(\{z_j\})$ (cf. [1], théorème 1.2 p. 444-445).

b) Des résultats du même type sont valables pour les équations hyperboliques du type

$$\begin{cases} \partial_t^2 u - \Delta u + f(u) + \varepsilon\partial_t u = g(x) \\ u|_{\partial\Omega} = 0 \end{cases}$$

Pour ces résultats, cf. [1], §6.

b) Dimension de Hausdorff de l'attracteur.

La théorie qui va suivre ne s'applique qu'aux semi-groupes compacts pour des raisons qui vont apparaître clairement. Par contre il n'est pas nécessaire que $\mathcal{M}(E)$ soit fini.

Rappelons que si X est un compact de E, la dimension de Hausdorff de X est définie de la façon suivante

- Soit $\mathcal{U}$ un recouvrement de X par une famille finie de boules $B_{r_i}(x_i)$ et $\delta \geq \operatorname{Sup}_i(r_i) = \delta(\mathcal{U})$.

On pose pour tous $N \in \mathbb{N}$, $\mathcal{U}$ et $\delta \geq \delta(\mathcal{U})$: $h_N(\mathcal{U},\delta,X) = \sum_i r_i^N$.

- Pour tout $\delta > 0$ on pose

$$h_N(\delta,X) = \inf_{\delta \geq \delta(\mathcal{U})} h_N(\mathcal{U},\delta,X)$$

- Enfin pour tout $N \in \mathbb{N}$, on considère

$$h_N(X) = \lim_{\delta \to 0} \operatorname{Sup} h_N(\delta,X).$$

La dimension de Hausdorff de X est alors

$$\dim(X) = \operatorname{Inf}\{N \in \mathbb{N},\ h_N(X) < +\infty\}$$

D'autre part si $L : H \to H$ est un *opérateur linéaire compact* et H un espace de Hilbert réel, on pose pour tout $n \in \mathbb{N}$

$$\hat{\mu}_n(L) = \operatorname*{Sup}_{V_n \in \mathcal{V}_n} \frac{\mu_n(L(B))}{\mu_n(B)}$$

où $\mathcal{V}_n$ est l'ensemble des sous-espaces vectoriels de H de dimension finie n et pour V_n fixé dans $\mathcal{V}_n$ on a représenté par

- B la boule de centre 0, de rayon 1 dans V_n
- μ_n la mesure n-dimensionnelle associée à un isomorphisme quelconque $V_n \approx \mathbb{R}^n$ obtenu en choisissant dans V_n une *base orthonormée* pour le produit scalaire induit dans V_n par celui de H.

On peut voir qu'en fait si on range dans l'ordre *décroissant* les valeurs propres $\beta_1, \beta_2, \ldots, \beta_n, \ldots$ de l'opérateur auto-adjoint compact$(LL^*)^{1/2}$, alors

$$\forall n \in \mathbb{N}, \quad \hat{\mu}_n(L) = \beta_1 \ldots \beta_n.$$

On remarque également que si L_1 et L_2 sont deux opérateurs linéaires compacts dans H, alors

$$\hat{\mu}_n(L_1 \circ L_2) \leq \hat{\mu}_n(L_1)\hat{\mu}_n(L_2).$$

Munis de ces diverses notations et définitions, nous pouvons énoncer le

Rappel 4.4. (Théorème de Douady -Oesterlé).

Soit H un Hilbert réel, $X \subset H$ un compact et S un opérateur : $H \to H$ tel que

$$\begin{cases} S \in C^1(U), \text{ U voisinage de X dans H} \\ S(X) = X \text{ et } S'(u) \text{ compact pour tout } u \in X. \end{cases}$$

Supposons qu'il existe $q < 1$ et $n \in \mathbb{N}$ tels que

$$\hat{\mu}_n(S'(u)) \leq q, \quad \forall u \in X.$$

Alors on a : $\dim X \leq n$.

Pour appliquer ce résultat aux attracteurs dans les équations aux dérivées partielles, on prendra $X = \mathcal{A}$ attracteur compact associé à l'équation

$$\partial_t u = A(u) \qquad u|_{t=0} = u_0$$

et on cherchera à utiliser le théorème de Donady-Oesterlé avec $S = S_t$ et t bien choisi.

Pour estimer $\hat{\mu}_n(S'(u))$ on considère alors l'*équation aux variations*

$$\partial_t v = A'(u(t))v = L(t)v.$$

Par exemple pour une équation de la chaleur non linéaire, L(t) est une perturbation de $-\Delta$. D'une façon générale, soit $\mathcal{Q} = \mathcal{Q}^*$ un opérateur (non borné) sur H et supposons que L(t) vérifie une inégalité de la forme :

$$(L(t)v,v) \leq (-\nu(t)\mathcal{Q}v + h(t)v,v), \quad v \in D(\mathcal{Q})$$

Sous des conditions techniques qui sont toujours satisfaites en pratiques, on en déduit (cf. [2]) que pour tout $n \in \mathbb{N}$ on a

$$\forall t \geq 0, \quad \hat{\mu}_n(G_t) \leq \exp\{-\mathrm{tr}_n\mathcal{Q}\int_0^t \nu(\tau)d\tau + n\int_0^t h(\tau)d\tau\}$$

où

$$\begin{cases} G_t = S_t'(u_0) \\ \mathrm{tr}_n\mathcal{Q} \text{ est la somme des } n \text{ "plus petites" valeurs propres de } \mathcal{Q}. \end{cases}$$

Si l'on fixe $t > 0$ et que l'on fait $n \to +\infty$, il est clair que puisque $\frac{\mathrm{tr}_n\mathcal{Q}}{n} \to +\infty$, le membre de gauche tend vers 0 dès que $\nu > 0$ et $h \in L^1(0,t)$. Il est donc clair que cette dernière inégalité implique, grâce au rappel 4.4, que $\dim(\mathcal{A}) < +\infty$.

En fait, cette méthode permet d'aller beaucoup plus loin et d'établir par exemple que pour l'équation

$$\begin{cases} \partial_t u = \nu\Delta u - f(u) + \lambda u - g(x) & (\nu > 0) \\ u|_{\partial\Omega} = 0 \text{ (ou } \frac{\partial u}{\partial n}|_{\partial\Omega} = 0) \end{cases}$$

on a lorsque $\lambda \to +\infty$

$$\dim(\mathcal{A}) \leq C(\frac{\lambda}{\nu})^{\frac{n}{2}}$$

Par une autre méthode basée sur la considération de la variété instable engendrée par les points stationnaires, on sait établir qu'en général on a aussi *dans le cas des conditions de Neumann*

$$\dim(\mathcal{A}) \geq c_1(\frac{\lambda}{1+\nu})^{\frac{n}{2}}, \quad c_1 > 0.$$

Enfin des résultats analogues sont possibles pour l'équation de Navier-Stokes en dimension 2. (cf. [2]). Les estimations de $\dim(\mathcal{A})$ dans ce cas font intervenir une inégalité spéciale portant sur le terme non-linéaire et dont la démonstration est analogue à celle de l'inégalité de Sobolev logarithmique de Brézis-Gallouët ([3]). Dans cette direction il faut noter les travaux anterieurs de Foias-Tema ([5]) et des résultats récents de Temam [12] et Ruelle [10].

Nous terminerons en remarquant que les problèmes suivants semblent actuellement ouverts

1) Minoration de dim($\mathcal{A}$) lorsqu'on considère l'équation de la chaleur non linéaire avec conditions aux limites de Dirichlet ?

2) Relation entre dim($\mathcal{A}$) et

$$\operatorname*{Max}_{y \in \mathcal{M}} \dim(M_+(\{y\})) \ ?$$

3) Cas des équations hyperboliques avec dissipation ? (on s'attend à des résultats semblables alors que la méthode précédente devient inapplicable).

REFERENCES.

[1] A.V. Babin, M.I. Vishik. Regular attractors of semigroups and evolution equations. J. Math. pures et appl., 62 (1983), 441-491.

[2] A.V. Babin, M.I. Vishik. Attracteurs de certaines équations d'évolution et estimations de leur dimension. Ouspekhi mat. naouk 38 (1983), 133-187. (en russe).

[3] H. Brezis, T. Gallouet. Nonlinear Schrödinger evolution equations. Nonlinear Analysis, T.M.A. 4 (1980), 667-671.

[4] A. Douady , J. Oesterle. Dimension de Hausdorff des attracteurs. C.R.A.S. Paris 290 (1980), 1135-1138.

[5] C. Foias, R. Temam . Some analytic and geometric properties of the solutions of the evolution Navier-Stokes equations. J. Math. Pures Appl., 58 (1979), 339-368.

[6] Ph. Hartman. *Ordinary differential equations*. John Wiley and Sons, New-York, London, Sydney (1964).

[7] I.C. Ilyashenko, A.N.Tchetaiev. Sur la dimension des attracteurs pour une classe de systèmes dissipatifs. Priklad-mat-meca, 46 (1982) 374-381.

[8] J. Mallet-Paret, Negatively invariant sets of compact maps and an extension of a theorem of Cartwight. J. Diff. Eq. 22 (1976), 331-348.

[9] J.E. Marsden, M. Mc-Cracken, *The Hopf bifurcation and its applications.* Springer-Verlag, New-York (1976).

[10] D. Ruelle, Large volume limit of distribution of characteristic exponents in turbulence. Comm. Math. Phys. 87 (1982), 287-302.

[11] S. Smale, Differentiable dynamical systems. Bull. Amer. Math. Soc., 73, 6 (1967), 747-817.

[12] R.Temam, Infinite dimensional dynamical systems in fluid mechanics, à paraître.

[13] R. Temam, Attractors for the Navier-Stokes equations. Dans ce volume.

M.I. VISHIK & A.V. BABIN

Institut des Problèmes de la Mécanique
de l'Académie des Sciences
101, rue Vernadskovo
117526 MOSCOU
URSS

C BARDOS & P DEGOND

Existence globale des solutions des équations de Vlasov–Poisson

I. INTRODUCTION.

Cet exposé est consacré à l'étude de la régularité et de l'unicité des solutions des équations de Vlasov-Poisson.

Les résultats obtenus et les méthodes utilisées font apparaître des similitudes importante avec l'étude de l'équation des ondes non linéaires.

Par exemple pour l'équation

$$\Box u + |u|^p u = 0 \text{ dans } \mathbb{R}_t \times \mathbb{R}^n \text{ , } u(x,0) = 0, \frac{\partial u}{\partial t}(x,0) = \Phi \qquad (1)$$

on sait (cf. Lions [8]) établir l'existence d'une solution faible en toute dimension n d'espace. Par contre on ne sait établir la régularité et l'unicité de la solution que dans le cas suivants :

(i) Φ est petit (d'autant plus petit que la dimension est grande $p \leq 2/(n-2)$).

(ii) La dimension d'espace est grande et p n'est pas trop petit (cette restriction disparait lorsque la dimension augmente mais pour n = 3 on doit savoir $p \geq 2$) et surtout les données initiales sont assez petites.

Le point (i) s'obtient en utilisant différentes formes du théorème de Sobolev pour considérer le terme non linéaire, tandis que le point (ii) s'obtient en utilisant les propriétés de dispersion de l'équation des ondes linéaires. De plus la dispersion permet de montrer qu'asymptotiquement (pour $|t| \to \infty$) les solutions de l'équation non linéaires se comportent comme des solutions de l'équation linéaire ce qui permet de développer une théorie du scottering non linéaire, dans cette direction les résultats les plus précis sont probablement ceux de Klainermann et Ponce [8].

On se propose donc de suivre la même démarche pour les équations de Vlasov-Poisson en rappelant des résultats du type "Sobolev" pour la dimension 2 et en introduisant une propriété de dispersion pour la dimension 3. Le problème ne présente bien sur pas d'intérêt physique en dimension supérieure à 3 mais cela vaut la peine de le décrire pour mettre en évidence le rôle de la dimension.

De plus la propriété de dispersion permet d'interpreter les résultats d'Ilner et Shinbrot [7] concernant l'existence de solution globale de l'équation de Boltzmann et de simplifier leur démonstration (cf. également Hamdache [6] pour une autre approche).

Les résultats concernant l'équation de Vlasov ont été annoncés dans une note C.R.A.S (Bardos-Degond [1]) et détaillés dans un article à paraître (Bardos-Degond [2]).

2. DESCRIPTION DES EQUATIONS DE VLASOV-POISSON ET DE LEURS INVARIANTS ELEMENTAIRES. EXISTENCE D'UNE SOLUTION FAIBLE.

Les équations de Vlasov-Poisson décrivent l'évolution d'un plasma dans lequel chaque type α $(1 \leq \alpha \leq N)$ de particule est décrit par une fonction de densité

$$f_\alpha(x,v,t) \geq 0$$

qui donne la quantité de particules qui au point x et à l'instant t sont animées de la vitesse v. Les collisions sont négligées et on suppose que le champ magnétique est petit par rapport à la vitesse de la lumière. On obtient alors le système suivant

$$\frac{\partial f_\alpha}{\partial t} + v.\nabla_x f_\alpha + \frac{q_\alpha}{m_\alpha} E.\nabla_v f_\alpha = 0 \qquad (1 \leq \alpha \leq N) \tag{2}$$

$$\nabla \wedge E = 0 \; ; \quad \nabla.E = \sum_\alpha 4 \cap q_\alpha \int f_\alpha(x,v,t)dv. \tag{3}$$

Dans (2) et (3) q_α et m_α désignent les charges et masses des particules de type α. (2) n'est autre qu'une loi de conservation , tandis que (3) est la loi de Coulomb (cf. Krall et Trivelpiece [9] ou Chen [5]).

L'équation (2) signifie que les fonctions $f_\alpha(x,v,t)$ sont constantes sur les trajectoires du champ

$$\dot{X} = V, \qquad \dot{V} = \frac{q_\alpha}{m_\alpha} E(X,t). \tag{4}$$

On en déduit que la positivité de $f\ (x,v,t)$ est préservée et que la norme de ces fonctions dans $L^\infty(\mathbb{R}^n_x \times \mathbb{R}^n_v)$ reste constante.

De plus le champ $(x,v) \to (v,E(x,t))$ est hamiltonien dans l'espace des phases $\mathbb{R}^n_x \times \mathbb{R}^n_v$; ainsi dans cet espace les normes $L^p(\mathbb{R}^n_x \times \mathbb{R}^n_v)$ sont préservées. On en déduit que la quantité

$$\rho(x,t) = \sum_\alpha 4\Pi \quad q_\alpha \int f_\alpha(x,v,t)dv \tag{5}$$

est uniformément bornée dans $L^1(\mathbb{R}^n_x)$.

On écrit alors le système (2), (3) sont la forme :

$$\frac{\partial f_\alpha}{\partial t} + \nabla_x(vf_\alpha) - \nabla_v.(\frac{q_\alpha}{m_\alpha} \nabla_x\Phi\ f_\alpha) = 0 \tag{6}$$

$$-\Delta\Phi = 4\Pi\rho = 4\Pi \sum_\alpha \int f_\alpha(x,v,t)dv.$$

Les estimations sur f dans L^∞ et sur ρ dans L^1 permettent alors de prouver l'existence d'une solution faible (cf. Arseneev[1]), en dimension n=1,2, ou 3.

Désormais, pour simplifier l'exposé on ne considérera qu'une seule fonction f(x,v,t), de signe variable solution du système :

$$\frac{\partial f}{\partial t} + v\nabla_x f - \nabla_x\Phi.\nabla_v f = 0 \tag{7}$$

$$-\Delta\Phi = \int f(x,v,t)dv, \tag{8}$$

et on supposera que pour t=0 on a :

$$\iint f(x,v,0)dxdv = 0. \tag{9}$$

La relation (9) remplace l'hypothèse selon laquelle les charges s'équilibrent à l'instant zéro :

$$\sum_{\alpha} \iint q_{\alpha} f_{\alpha}(x,v,0)d\ dv = 0\ ; \tag{10}$$

Elle est invariante au cours du temps. Tous les résultats mathématiques décrits sur le modèle (7),(8) se transposent aisément aux vraies équations.

3. EXISTENCE D'UNE SOLUTION FORTE POUR TOUT TEMPS EN DIMENSION DEUX ET POUR UN TEMPS PETIT EN DIMENSION 3 SELON UKAI ET OKABE.

On sait déjà que l'on a :

$$\int |\rho(x,t)|dx \qquad \iint |f(x,v,t)|dxdv = cte \tag{11}$$

ce qui a été utilisé pour prouver l'existence d'une solution faible ; pour obtenir une solution régulière il serait nécessaire de disposer d'estimations à priori sur la norme de ρ dans $L^p(\mathbb{R}^n_x)$ $(p > 1)$; ou,ce qui revient au même, de montrer qu'il n'y a pas trop de particules qui prennent de grandes vitesses, de manière à controler l'expression

$$\int (\int f(x,v,t)dv)^p dx.$$

Pour effectuer ce controle on dispose du théorème des accroissements finies sur les solutions du système différentiel :

$$\dot{X} = V \qquad \dot{V} = E(X,t). \tag{12}$$

Ainsi une estimation a priori sur ρ est naturellement associée à une estimation à priori sur E. On utilise alors le lemme (de type inégalité de Sobolev) suivant.

<u>Lemme</u> 1. *Il existe une constante* c_n *ne dépendant que de la dimension de l'espace telle que pour toute fonction* Φ *régulière, tendant vers zéro à l'infini, on ait :*

$$|\nabla\Phi|_\infty \le c_n |\Delta\Phi|_1^{1/n} \, |\Delta\Phi|_\infty^{1-1/n} \; . \tag{13}$$

On a alors dans $\mathbb{R}^n$ $(n \ge 2)$ ([1]) l'estimation suivante :

<u>Théorème</u> 1. Ukai-Okabe [10].

Soit $f(x,v,t)$ *une solution régulière de l'équation de Vlasov Alors :*

$$M(t) = \sup_x |E(x,t)| = |\nabla\Phi(.,t)|_\infty .$$

vérifie l'inégalité suivante :

$$M(t) \le C\left(1+\left(\int_0^t M(s)ds\right)^n\right)^{(n-1)/n} \; . \tag{14}$$

où C *désigne une constante indépendante de* t *mais dépendant des quantités*

$$\sup_{x,v} |f(x,v,0)| \, , \; \iint |f(x,v,0)| dv dx$$

$$\sup_{x,v} \int_{|\xi|>|v|/2} |f(x,\xi,0)| d\xi$$

(qui sont supposées finies!).

<u>Remarque</u> 1. L'inégalité (14) va être établie pour des solutions régulières ; mais il s'agit d'une majoration a priori qui permet en fait de prouver l'existence de ces solutions régulières. Lorsque n=2 (14) est en fait l'inégalité de Gronwall classique ce qui donne un résultat valable pour tout $t > 0$. Par contre pour $n > 2$ on a une inégalité non linéaire ce qui ne conduit qu'à un résultat valable pour un temps fini.

([1]) On note par $|.|_p$ les normes dans $L^p(\mathbb{R}^n_x)$.

Démonstration du Théorème 1. Compte tenu du lemme 1 on a :

$$\begin{aligned} M(t) &\leq C|\rho(.,t)|_1^{1/n}|\rho(.,t)|_\infty^{1-1/n} \\ &\leq C|\rho(.,t)|_\infty^{1-1/n} \end{aligned} \tag{15}$$

(car la norme de ρ dans L^1 est conservée).

Comme f est constant le long des trajectoires du champ (v,E) il convient d'introduire la solution X(t,s,x,v), V(t,s,x,v) du système

$$\frac{\partial X}{\partial s} = V, \quad \frac{\partial V}{\partial s} = E(X,t) \ ; \tag{16}$$

$$X(t,t,x,v) = x, \ V(t,t,x,v) = v.$$

En posant :

$$X(t,0,x,v) = X, \quad V(t,0,x,v) = V$$

on obtient :

$$\begin{aligned} |\rho(x,t)| &\leq \int |f(x,v,t)|dv = \int |f(X,V,0)|dv \\ &\leq \int_{|v|\leq 2\int_0^t M(s)ds} |f(x,v,0)|dv + \int_{|v|\geq 2\int_0^t M(s)ds} |f(X,V,0)|dv. \end{aligned} \tag{17}$$

Comme la donnée initiale est dans L^∞ le premier terme du dernier membre de (17) se majore par

$$C(\int_0^t M(s)ds)^n.$$

D'après le théorème des accroissements finis on a :

$$|v-V| \leq \int_0^t M(s)ds. \tag{18}$$

On en déduit que pour $|v| > 2 \int_0^t M(s)ds$, on a :

$$|V| \geq |v| - \int_0^t M(s)ds \geq |v|/_2. \tag{19}$$

Ainsi le second terme du dernier membre de (17) est majoré par l'expression

$$\sup_{x,v} \int_{|\xi|>|v|/2} |f(x,\xi,0)|d\xi$$

qui est finie et indépendante de t. Ainsi on a :

$$|\rho(.,t)|_\infty \leq C(1+(\int_0^t M(s)ds)^n), \tag{20}$$

ce qui avec (15) donne la relation (14).

Comme on l'a souligné, cette méthode ne donne en dimension 3 qu'un résultat valable pour un temps fini. Pour obtenir des résultats globaux il convient de mettre en évidence une propriété de dispersion. L'équation linéarisé au voisinage de $f = \Phi = 0$ n'est autre que l'équation de transport :

$$\frac{\partial f}{\partial t} + v.\nabla f = 0 \tag{21}$$

on en déduit que sa solution est donnée par

$$f(x,v,t) = f(x-vt,v;0) \tag{22}$$

Ainsi si on suppose que la fonction

$$h(x) = \sup f(x,v,0)$$

est intégrable on obtient, pour la charge correspondante, la majoration suivante :

$$
\begin{aligned}
|\rho(x,t)| &\leq \int |f(x-vt,v,0)|dv \\
&\leq \int h(x-vt)dv \\
&= \int h(X) \left|\left(\frac{\partial X}{\partial v}\right)_t\right|^{-1} dX
\end{aligned}
\tag{23}
$$

ou $|(\frac{\partial X}{\partial v})_t|$ désigne le jacobien de l'isomorphisme de $\mathbb{R}^n$ dans $\mathbb{R}^n$, défini par

$$v \to X_t(v) = x - vt. \tag{24}$$

De (21) on déduit donc la relation de dispersion

$$|\rho(.,t)|_\infty \leq \frac{1}{t^n} \int h(X)dX \tag{25}$$

ce qui prouve que la charge s'étale d'autant plus vite que la dimension est élevée.

Cette remarque va permettre de construire, par exemple par itération des solutions régulières en dimension d > 3. Nous traitons d'abord le cas d > 3.

4. CONSTRUCTION D'UNE SOLUTION REGULIERE EN DIMENSION SUPERIEURE A 3.

On utilise une méthode itérative ; pour $\rho_n(x,t)$ donné dans $\mathbb{R}^d_x \times \mathbb{R}^+_t$, on calcule $E_n(x,t)$ par la formule :

$$-\Delta\Phi_n = \rho_n \ ; \quad E_n = - \nabla\Phi_n \tag{26}$$

puis on résout l'équation de transport linéaire :

$$
\left\{
\begin{aligned}
&\frac{\partial f_{n+1}}{\partial t} + v.\nabla_x f_{n+1} + E_n.\nabla_v f_{n+1} = 0, \\
&\qquad f_{n+1}(x,v,0) = f_o(x,v).
\end{aligned}
\right.
\tag{27}
$$

Enfin on calcule $\rho_{n+1}(x,t)$ par la formule :

$$\rho_{n+1}(x,t) = \int f_{n+1}(x,v,t)dv \tag{28}$$

et on a le

Théorème 2. *On suppose que la donnée initiale* $f_0(x,v)$ *est assez régulière ; alors si l'expression*

$$I(1)=\int_{v,|\xi|\leq 1} |f_0(x+\xi,v)|dx \tag{29}$$

est assez petite, $\rho_n(x,t)$ *vérifie la majoration à priori suivante :*

$$|\rho_n(x,t)| \leq C/(1+t)^d \tag{30}$$

où C *désigne une constante ne dépendant que des données initiales, en particulier indépendante de* t *et de* n.

Remarque 2. On déduit facilement de la relation (30) l'existence d'une solution régulière $f(x,v,t)$ de l'équation

$$\left\{\begin{array}{l} \frac{\partial f}{\partial t} + v.\nabla_x f + E\nabla f = 0 \\ E = -\nabla\Phi \ , \ -\Delta\Phi = \int f(x,v,t)dx \\ f(x,v,0) = f_0(x,v) \end{array}\right. \tag{31}$$

cette solution vérifie en particulier la majoration

$$|\rho(x,t)| = |\int f(x,v,t)dv| \leq \frac{C}{(1+t)^d} . \tag{32}$$

Démonstration du théorème 2. On suppose que (30) est vérifiée à l'ordre n avec C assez petit et on va en déduire que cette relation est encore vraie à l'ordre n+1, avec la même constante C.

Il est d'abord facile à vérifier, par un lemme de Gronwall non linéaire (lié à l'existence locale de la solution) que l'on a pour $0 \le t \le T^*$ (T^* fini dépendant de la donnée initiale),

$$|\rho_n(x,t)| \le C_1,$$

il suffit donc de prouver que si la relation

$$|\rho_n(x,t)| \le C/(1+t)^d \tag{33}$$

est vraie à l'ordre n, elle est vraie, pour $t \ge T^*$ à l'ordre n+1. Pour cela on va introduire les trajectoires $(X_n(t,s,x,v),V_n(t,s,x,v))$ de l'équation différentielle

$$\left.\begin{aligned} &\dot{X}_n = V_n,\ \dot{V}_n = E_n(X_n,s)\ ; \\ &X_n(t,t,x,v) = x,\ V_n(t,t,x,v) = v \end{aligned}\right. \tag{34}$$

et on va mesurer la déviation de ces trajectoires par rapport à la trajectoire libre

$$(X = x - (t-s)v,\ V = v).$$

Comme on a $\ddot{X}_n(s) = E_n(X_n(s),s)$, on peut écrire :

$$\begin{aligned} X_n(0) &= x - (t\dot{X}_n(t) + \int_0^t (t-s)\ddot{X}_n(t-s))ds \qquad (35) \\ &= x - tv + \int_0^t (t-s)E_n(X_n(t-s),t-s)ds \end{aligned}$$

on en déduit la relation :

$$|X_n(0) - (x-tv)| \le \int_0^t s|E_n(X_n(s),s)|ds. \tag{36}$$

On utilise le lemme 1

$$|E_n(.,s)|_\infty \le C_1 C |\rho_n(.,s)|_1^{1/d} |\rho_n(.,s)|_\infty^{(d-1)/d} \tag{37}$$

comme la norme dans L^1 de $\rho_n(.,s)$ est conservée, on peut déduire de (37) et de l'hypothèse de récurrence (33), la relation

$$\begin{aligned} |X_n(0) - (x-tv)| &\le C_2 C\left(\int_0^t \frac{s}{(1+s)^{(d-1)}}\, ds\right) \\ &\le C_3 C \int_0^\infty \frac{ds}{(1+s)^{(d-2)}} . \end{aligned} \tag{38}$$

Pour d > 3 cette dernière intégrale est convergente et si la constante C figurant dans (33) est assez petite, on a :

$$|X_n - (x-tv)| \le 1. \tag{39}$$

on en déduit la majoration

$$\begin{aligned} |\rho_{n+1}(x,t)| &\le \int |f_0(X_n(0), V_n(0))|dv \\ &\le \int |f_0(X_n(0) - (x-tv) + x - tv, v)|dv \\ &\le \int \sup_{|\xi|\le 1,} |f_0(x-tv+\xi, v)|dv \\ &\le \left(\int \sup_{v,|\xi|\le 1} |f_0(X+\xi, v)|dX\right)\frac{1}{t^d} . \end{aligned} \tag{40}$$

Finalement on déduit de (40) la relation :

$$|\rho_{n+1}(.,t)| \le I(1)\, \frac{1}{t^d} , \tag{41}$$

ce qui implique (33) à l'ordre (n+1) pour $t > T^*$ pourvu que I(1) soit assez petite.

La majoration faisant intervenir des quantités du type I(1) est en fait grossière et pour traiter le cas de la dimension 3 il convient d'estimer précisément le jacobien de la transformation

$$v \to X(t,0,x,v)$$

c'est cette demarche qui va être utilisée au paragraphe suivant.

5. LE CAS DE LA DIMENSION 3.

On commence par prouver un lemme sur les systèmes différentiels.

Lemme 2. *On suppose que* $(x,t) \to E(x,t)$ *est un champ deux fois continûment différentiable sur* $\mathbb{R}_x^3 \times \mathbb{R}_t$ *satisfaisant les estimations*

$$\|\nabla_x E(.,t)\|_\infty \leq \eta/(1+t)^{5/2} \tag{42}$$

$$\|\nabla_x^2 E(.,t)\|_\infty \leq \eta/(1+t)^{5/2} \quad ({}^1) \tag{43}$$

avec $\eta < 1$ *alors les matrices*

$$\frac{\partial V}{\partial v}(t,s,x,v),\ \frac{\partial V}{\partial x}(t,s,x,v),\ \frac{\partial X}{\partial x},\ \frac{\partial X}{\partial v}$$

et les tenseurs

$$\frac{\partial^2 X}{\partial x^2}(t,s,x,v),\ \frac{\partial^2 V}{\partial x^2}(t,s,x,v)$$

vérifient en tout point $s \in [0,t]$ *les estimations suivantes*

$$\left|\frac{\partial X}{\partial v} - (t-s)Id\right| + \left|\frac{\partial V}{\partial v}\right| \leq C\eta(t-s) \tag{44}$$

([1]) Dans l'énoncé de ce lemme on peut remplacer 5/2 par tout nombre supérieur à 2. Dans la suite il faudra considérer un exposant strictement compris entre 2 et 3 ; il est plus simple de fixer la valeur 2.5 dès maintenant.

$$|\frac{\partial X}{\partial x} - Id| + |\frac{\partial V}{\partial x}| \leq C\eta. \qquad (45)$$

$$|\frac{\partial^2 X}{\partial x^2}| + |\frac{\partial^2 V}{\partial v^2}| \leq C\eta.$$

Remarque 3. Toutes ces estimations sont trivialement vérifiées lorsque $E \equiv 0$ car alors X et V sont donnés par les relations :

$$X(s) = x - (t-s)v, \quad V(s) = v.$$

Démonstration. La démonstration se fait en remarquant que $\frac{\partial X}{\partial v}$ et $\frac{\partial X}{\partial x}$ sont solutions des équations différentielles du second ordre :

$$\frac{\partial \ddot{X}}{\partial v} = \nabla_x E. \frac{\partial X}{\partial v}, \quad \frac{\partial X}{\partial v}(t) = 0 \quad \frac{\partial \dot{X}}{\partial v} = I \qquad (47)$$

$$\frac{\partial \ddot{X}}{\partial x} = \nabla_x E \frac{\partial X}{\partial n}, \quad \frac{\partial X}{\partial x}(t) = I, \quad \frac{\partial \dot{X}}{\partial x} = 0 \qquad (48)$$

et en utilisant ensuite la formule de Taylor au second ordre avec reste intégral, comme dans la IIe partie.

Corollaire 1. *Sous les hypothèses du lemme 2 et pour* $\eta > 0$ *assez petit les assertions suivantes sont vraies.*

(i) *Le déterminant de* $\frac{\partial X}{\partial v}$ *vérifie la minoration*

$$|\det(\frac{\partial X}{\partial v})| \geq \frac{1}{2}(t-s)^3. \qquad (49)$$

(ii) *l'application* $v \to \frac{\partial X}{\partial v}(t,s,x,v)$ *est toujours injective.*

Démonstration. On écrit :

$$\frac{\partial X}{\partial v} = (t-s)I + (t-s)B(s) = (t-s)(I+B(s)). \qquad (50)$$

D'après (44) B(s) est une matrice de norme inférieure à $C\eta$ donc pour η assez petit on a

$$|\det(I+B(s))| \geq \frac{1}{2}$$

ce qui prouve (i).

Le point (ii) se prouve en considérant la sous-variété de dimension 3, Γ_s définie dans $\mathbb{R}^3_x \times \mathbb{R}^3_v \times \mathbb{R}_t$ par la formule

$$\Gamma_s = \{(X,V) \mid X = X(t,s,x,v),\ V = V(t,s,x,v),\ v \in \mathbb{R}^3\}. \qquad (51)$$

Pour s assez proche de t la projection de Γ_s sur $\mathbb{R}^3_x$ est injective si sur l'intervalle $[0,t]$ cette propriété cessait d'être vraie on introduirait le point $s^* > 0$ borne inférieure des $\tilde{s}$ tels que pour $s \in]\tilde{s},t]$, la projection de Γ_s sur $\mathbb{R}^3_x$ soit injective. Par un argument de transversalité, on montre alors qu'il existerait v^* tel que l'on ait :

$$\det \frac{\partial X}{\partial v}(t,s^*,x,v^*) = 0$$

ce qui contredit le point (i).

<u>Remarque</u> 4. Le corollaire 1 signifie que sous les hypothèses du lemme 1 les trajectoires du système hamiltonien n'engendrent pas de caustiques. Ceci est du au fait suivant : pour s voisin de t le champ E est près de zero d'ordre η et pour (t-s) grand il s'écrase en $1/(t-s)^{2.5}$. Dans le théorème 3 on va donc montrer l'existence d'une solution dont les particules n'engendrent pas de caustique, et inversement l'apparition éventuelle de singularités dans la solution est peut être liée à l'apparition de caustique dans les trajectoires.

La nécessité d'estimer les dérivées premières et secondes dans $L^\infty(\mathbb{R}^3_x)$ de E(x,t) conduit à établir le lemme suivant.

Lemme 3. *Pour une fonction* ϕ *tendant vers zéro à l'infini solution dans* $\mathbb{R}^3$ *de l'équation de Poisson :*

$$-\Delta\phi = \rho, \text{ avec } \rho \in L^1(\mathbb{R}^n) \cap W^{\phi,\infty}(\mathbb{R}^n)$$

les estimations suivantes sont valables.

$$|\phi|_\infty \leq C|\rho|_\infty^{1/3}|\rho|_1^{2/3} \tag{52}$$

$$|\nabla\phi|_\infty \leq C|\rho|_\infty^{2/3}|\rho|_1^{1/3} \tag{53}$$

$$|D^2\phi|_\infty \leq C_\theta|\rho|_\infty^{3(1-\theta)/(3+\theta)}|\nabla\phi|_\infty^{3\theta/(3+\theta)}|\rho|^{\theta/(3+\theta)} \tag{54}$$

$$\forall\theta \in]0,1[.$$

Dans (54) C_θ désigne une constante dépendant de θ et $D^2\phi$ n'importe quelle dérivée seconde.

Ces estimations sont classiques; la moins triviale (54) est obtenue en utilisant l'estimation holderienne

$$|D^2\phi|_\infty \leq |D^2\phi|_{0,\theta} \leq C_\theta|\rho|_{0,\theta}^{3/(3+\theta)}|\rho|_1^{\theta/(3+\theta)} \tag{55}$$

et en interpolant grossièrement l'espace $C_{0,\theta}$ entre les espace L^∞ et $W^{1,\infty}$.

Les outils ci-dessus permettent maintenant de prouver facilement le

Théorème 3. *On suppose que la donnée initiale* $f(x,v,0) = f_0(x,v)$ *est continument différentiable dans* $\mathbb{R}_x^3 \times \mathbb{R}_v^3$ *et satisfait les estimations suivantes*

$$|f_0(x,v)| \leq \varepsilon/(1+|x|)^4(1+|v|)^4 \tag{56}$$

$$|\nabla_{x,v} f_0(x,v)| + |D_{x,v}^2 f_0(x,v)| \leq \varepsilon/(1+|x|)^4(1+|v|)^4. \tag{57}$$

alors il existe $\varepsilon_0 > 0$ *tel que pour* $\varepsilon < \varepsilon_0$ *l'équation de Vlasov-Poisson*

$$\frac{\partial f}{\partial t} + v.\nabla_x f + E.\nabla_v f = 0 \tag{58}$$

$$-\Delta\Phi = \rho, \ E = - \nabla\Phi \tag{59}$$

ou ce qui revient au même

$$E = \frac{1}{4\pi} \int_{\mathbb{R}^3} \frac{x-y}{|x-y|^3} \rho(y,t)dy \tag{60}$$

admet une unique solution globale continument différentiable en (x,v,t) *qui satisfait de plus l'estimation uniforme :*

$$|\rho(.,t)|_1 + |\nabla_x \rho(.,t)|_1 \leq A\varepsilon \tag{61}$$

et l'estimation de dispersion

$$|\rho(.,t)|_\infty + |\nabla_x f(.,t)|_\infty + |\nabla_x^2 \rho(.,t)|_\infty \ \leq \ A/(1+t)^3 \tag{62}$$

où A *désigne une constante indépendante de* ε.

<u>Démonstration</u>. La démonstration se fait par itération en introduisant la suite $\rho_n.f_n$ définie par

$$E_n(x,t) = \frac{1}{4\pi} \int \frac{x-y}{|x-y|^3} \rho_n(y,t)dy \tag{63}$$

$$\begin{gathered} \frac{\partial f_{n+1}}{\partial t} + v\nabla_x f_{n+1} + E_n \nabla_x f_{n+1} = 0, \\ f_{n+1}(x,v,0) = f_0(x,v), \end{gathered} \tag{64}$$

et en montrant que les itérés vérifient tous la relation (62). En effet en utilisant le résultat d'existence locale d'Ukai et Okabe, on montre que cette propriété est vraie pour tout n sur un intervalle $[0,T^*[$. Ensuite on suppose que cette relation est vraie à l'ordre n ; on en déduit en utilisant le lemme 3 avec $\theta = 3/5$ que l'on a :

$$|\nabla E_n(.,t)|_\infty + |D^2E_n(.,t)|_\infty \leq CA\varepsilon^{1/6}/(1+t)^{2.5}. \tag{65}$$

Il en résulte que si on a $CA\varepsilon^{1/6} < 1$ on peut appliquer aux trajectoires du champ

$$\dot{X} = V,\ \dot{V} = E_n(X,V),\ X(t,t,x,v) = x,\ V(t,t,x,v) = v$$

les résultats du lemme 2. On a ainsi :

$$\begin{aligned}|\rho_{n+1}(x,t)| &\leq \int |f_{n+1}(x,v,t)|dv \\ &\leq \int |f_0(X_n(t,0,x,v),\ V_n(t,0,x,v)|dv \\ &\leq \int \varepsilon \frac{1}{(1+|X_n(t,0,x,v)|)^4} dv\end{aligned} \tag{66}$$

(d'après l'hypothèse (56)).

Dans le dernier terme de (66) on effectue le changement de variable

$$v \to X = X_n(t,0,x,v)$$

et on obtient (corollaire 1)

$$|\rho_{n+1}(x,t)| \leq \varepsilon \int \frac{1}{(1+|X|^4)} |(\frac{\partial v}{\partial X})| dX \tag{67}$$

soit toujours d'après le corollaire 1,

$$|\rho_{n+1}(x,t)| \leq \frac{\varepsilon}{2t^3} \int \frac{dX}{(1+|X|)^4}\ . \tag{68}$$

Il suffit ensuite de remarquer que si ε est assez petit on a alors pour $t > T^*$.

$$|\rho_{n+1}(.,t)|_\infty \leq A/(1+t)^3.$$

Les autres termes du premier membre de (62) calculés à l'ordre n+1, se majorent de manièrent analogue, car on a par exemple

$$\nabla\phi_{n+1} = \int \left(\frac{\partial f}{\partial x}(X,V)\frac{\partial X}{\partial x} + \frac{\partial f}{\partial V}\cdot\frac{\partial V}{\partial x}\right)dv$$

et on utilise la majoration uniforme de $\frac{\partial X}{\partial x}$ et $\frac{\partial V}{\partial x}$ donnée par (45). Pour les dérivées secondes on utiliserait de même (46).

On a donc prouvé que si la relation (62) est vraie à l'ordre n, elle est vraie à l'ordre (n+1) et avec la même constante A, pourvu que A soit assez petit. On termine aisément la démonstration.

Remarque 5. Les trajectoires $X_n(x,v,t)$ ne possèdent pas de caustiques et cette propriété passe facilement à la limite.

5. COMPARAISON AVEC LA SOLUTION DE L'EQUATION DE BOLTZMANN DONNEE PAR ILNER ET SHINBROT.

Ilner et Shinbrot [7] ont récemment montre que l'équation de Boltzmann admettait une solution globale pourvu que la donnée initiale soit assez petite par rapport au libre parcours moyen ε. Contrairement à la théorie usuelle (Nishida [14] ou Caflish [4]) Ilner et Shinbrot considérent des données initiales voisines du vide et non d'une maxwellienne. Cette démonstration a été reprise et généralisée par Hamdache [6] selon des idées développées par Tartar [13] sur le modèle de Broadwell ; mais on peut remarquer qu'en fait la démonstration repose sur une notion de dispersion analogue à celle utilisée ci-dessus, ce qui conduit à un énoncé très simple que l'on va résumer ici.

On considère donc l'équation de Boltzamnn

$$\frac{\partial f}{\partial t} + v.\nabla f = \frac{1}{\varepsilon}Q(f,f). \tag{69}$$

où Q(f,f) désigne l'opérateur de collision calculé dans le cas des potentiels maxwelliens.

$$Q(f,f) = \iint_{\mathbb{R}^3_v\times S^2}(f(v')f(v_1)-f(v)f(v_1))q(|v-v_1|,\omega)d\omega\ .\ dv_1. \tag{70}$$

Dans (70) v' et v_1' sont déterminés par v, v_1 et ω à l'aide des formules

$$v' = v - (\omega,(v-v_1))\omega, \quad v_1' = v_1 + (\omega,(v-v_1))\omega. \tag{71}$$

qui conduisent à la conservation du moment et de l'énergie cinétique :

$$v' + v_1' = v + v_1, \quad |v'|^2 + |v_1'|^2 = |v|^2 + |v_1|^2. \tag{72}$$

On suppose (pour simplifier cf Hamdache [6] pour d'autres cas) que q est une fonction uniformément bornée. On utilise les notations suivantes :

$$f' = f(v'), \quad f_1' = f(v_1'). \tag{73}$$

On a alors le

<u>Théorème</u> 4. *On suppose que la donnée initiale* $f(x,v,0) = f_0(x,v)$ *est continument dérivable et vérifie la majoration :*

$$0 \le f_0(x,v) \le Ae^{-|x|^2} \tag{74}$$

avec A *assez petit (en particulier par rapport à* ε*).*

Alors l'équation de Boltzmann

$$\frac{\partial f}{\partial t} + v.\nabla_x f = \frac{1}{\varepsilon}Q(f,f), \quad f(x,v,t) = f_0(x,v) \tag{75}$$

admet une unique solution, définie dans $\mathbb{R}_t \times \mathbb{R}^3 \times \mathbb{R}^3$ *continument différentiable, satisfaisant la majoration à priori :*

$$0 \le f(x,v,t) \le Be^{-|x-vt|^2} \tag{76}$$

Asymptotiquement cette solution se comporte comme la solution d'une équation de Transport. Plus précisément il existe deux fonctions $g_{\pm}$ *solutions de l'équation de Transport :*

$$\frac{\partial g_{\pm}}{\partial t} + v.\nabla g_{\pm} = 0 \tag{77}$$

telles que l'on ait :

$$\lim_{t\to\pm\infty} \sup_{x,v} (|f(x,v,t) - e^{-|x-vt|^2} g_{\pm}(x,v,t)|) = 0 \tag{78}$$

<u>Démonstration</u>. Encore une fois le point essentiel est la preuve d'une estimation à priori le reste se fait ensuite aisément. On pose donc

$$f(x,v,t) = e^{-|x-vt|^2} g(x,v,t)$$

et on remarque que l'on a :

$$(\frac{\partial}{\partial t}+v.\nabla)(e^{-|x-vt|^2}) = 0. \tag{79}$$

En reportant dans (75) on obtient donc la solution

$$e^{-|x-vt|^2}(\frac{\partial g}{\partial t}+v\nabla g) = \frac{1}{\varepsilon}\iint(e^{-|x-v_1't|^2-|x-v't|^2}g'g_1' - e^{-|x-vt|^2-|x-v_1t|^2}gg_1) \times \times q(|v-v_1|,\omega)dv_1 d\omega. \tag{80}$$

Comme on a à cause de (72)

$$|x-vt|^2 - |x-v_1't|^2 - |x-v't|^2 = - |x-v_1t|^2 \tag{81}$$

l'équation (80) se simplifie pour donner :

$$\frac{\partial g}{\partial t} + v.\nabla g = \frac{1}{\varepsilon}\iint e^{-|x-v_1t|^2}(g'g_1'-gg_1)q dv_1 d\omega. \tag{82}$$

L'équation de Transport est un groupe unitaire dans $L^{\infty}(\mathbb{R}_x^3 \times \mathbb{R}_v^3)$ ainsi en posant $X(t) = \sup_{x,v} g(x,v,t)$, on déduit de (82) la relation

$$\frac{d}{dt_+} X(t) \le \frac{2C}{\varepsilon} \iint e^{-|x-v_1t|^2} (X(t))^2 dv_1 d\omega \tag{83}$$

(La constance C provient de la fonction q). On intégre par rapport à ω le second membre de (83) et en changeant la constante on obtient

$$\frac{d}{dt_+} X(t) \le \frac{C}{\varepsilon}(X(t))^2 \int e^{-|x-v_1t|^2} dv_1 \quad (^1) \tag{84}$$

Le changement de variable de dispersion $X = x - v_1 t$ donne alors

$$\frac{d}{dt_+} X(t) \le \frac{C}{\varepsilon}(X(t))^2 \frac{1}{t^3}. \tag{85}$$

Soit en utilisant uniquement une majoration locale :

$$\frac{d}{dt_+} X(t) \le \frac{C}{\varepsilon} \frac{1}{(1+t)^3} (X(t))^2. \tag{86}$$

C'est-à-dire

$$X(t) \le \frac{X(0)}{1-\frac{C}{\varepsilon}(\int_0^t \frac{ds}{(1+s)^3})X(0)} \tag{87}$$

On a ainsi une majoration globale dès que la relation

$$\frac{C}{\varepsilon} \int_0^\infty \frac{ds}{(1+s)^3} X(0) < 1 \tag{88}$$

est vérifier, ce qui permet de prouver l'existence d'une solution de (75) vérifiant (76).

Pour établir le comportement asymptotique, on remarque que g est solution de l'équation

---

(1) On considère le cas $t \ge 0$ (le cas $t \le 0$ est analogue) et on désigne par $\frac{d}{dt_+}$ la dérivée à droite.

$$\frac{\partial g}{\partial t} + v.\nabla_x g = \frac{1}{\varepsilon}\iint e^{-|x-v_1 t|}(g'g_1'-gg_1)q dv.d\omega \tag{89}$$

$$= \frac{1}{\varepsilon} R(t,x,v)$$

ou $R(t,x,v)$ vérifie la majoration uniforme

$$\left|\frac{1}{\varepsilon} R(t,x,v)\right| \leq \frac{1}{\varepsilon}\frac{C}{(1+t)^3}. \tag{90}$$

On désigne ensuite par A l'opérateur d'advection $-v\nabla_x$ qui est générateur d'un groupe unitaire dans $L^\infty(\mathbb{R}_x^3 \times \mathbb{R}_v^3)$ et on va montrer qu'il existe g_+ tel que l'on ait

$$\lim_{t\to+\infty} |e^{tA}g_+ - g(x,v,t)|_\infty = 0. \tag{91}$$

Comme e^{tA} est un groupe cela revient à prouver que $e^{-tA}g$ converge dans $L^\infty(\mathbb{R}_x^3 \times \mathbb{R}_v^3)$ ou que l'on a :

$$\int_{t_1}^{t_2} \left|\frac{d}{ds}(e^{-sA}g)\right|_\infty ds \leq \eta \tag{92}$$

pourvu que t_1 et t_2 soient assez grand.

Maintenant on écrit :

$$\left|\frac{d}{ds}(e^{-sA}g)\right|_\infty = \left|e^{-sA}(\frac{\partial g}{\partial t} - Ag)\right|_\infty$$

$$= \left|e^{-sA}(\frac{\partial g}{\partial t} + v\nabla_x g)\right|_\infty \tag{93}$$

$$= \left|\frac{\partial g}{\partial t} + v.\nabla_x g\right|_\infty \leq \frac{C}{\varepsilon(1+s)^3}.$$

Comme ce dernier terme est intégrable, on a prouvé l'existence de g_+. On a donc :

$$\lim_{t\to\infty} |f(x,v,t) - e^{-|x-vt|^2}g_+(x-vt,v)|_\infty = 0, \tag{94}$$

ce qui prouve la deuxième assertion du théorème.

6. REMARQUES DIVERSES ET PROBLEMES OUVERTS.

1. Au paragraphe 3 nous avons montré que la charge se disperse comme $1/t^3$ mais nous n'avons pas été capable de faire une description aussi complète du comportement asymptotique de la solution des équations de Vlasov, le terme $|E.\nabla_v f|_\infty$ ne semble pas être intégrable par rapport à t (nos estimations le donnent en $1/_t$).

2. En dimension 1 et 2 on dispose de bons théorèmes d'existence mais il n'existe pas de résultat mathématique précis sur le comportement asymptotique des solutions.

3. Un résultat semblable a celui du paragraphe 3 devrait être valable pour le système complet (faisant intervenir également le champ magnétique B) dit de Vlasov-Maxwell. L'esprit de la démonstration semble analogue, mais il apparait de nombreuses difficultés techniques. La démonstration d'un résultat d'existence locale pour ce système n'est d'ailleurs pas évidente elle ne date que de 83 Wollmann [10](ou Degond [13]).

4. Les équations du type Vlasov (Maxwell ou Poisson) semblent, plus que l'équation de Boltzmann concernées par des problèmes aux limites. Il serait intéressant de déterminer quelles sont les conditions aux limites convenables et traiter les problèmes correspondants. Les résultats du paragraphe 2 doivent facilement s'adapter au cas d'un domaine borné (en espace bien sur, les vitesses sont toujours dans $\mathbb{R}^n$) par contre ceux du paragraphe 3 doivent s'adapter au problème extérieur.

REFERENCES.

[1] A.A. Arseneev, Global existence of a weak solution of Vlasov system of equations. U.S.S.R. Comput. Math. and Math. Phys. 15 (1975), 131-143.

[2] C. Bardos et P. Degond, Global existence for the Vlasov-Poisson equation in Espace variables with small initial data (a paraitre au Journal of Non Linear Functional Analysis).

[3] C. Bardos et P. Degond, Existence globale et comportement asymptotique de la solution de l'équation de Vlasov-Poisson Note C.R.A.S.

série A tome 297 (321-323) (1983).

[4] R. Caflish, The fluid dynamical limit of the non linear. Boltzmann equation. Comm. in Pure and Appl. Math. 33 651-666 (1980).

[5] F. Chen, Introduction to plasma physic Plenum.

[6] K. Hamdache, Quelques résultats pour l'équation de Boltzmann Note C.R.A.S. t. 299 1 n°10, 1984 pp. 431-434.

[7] R. Ilner et M. Shinbrot, à paraître.

[8] S. Klainermann et G. Ponce, Global Small Amplitude solutions to non linear evolution equation. Com. Pure Appl. Math. 36. 1 (1983) 133-141.

[9] N. Krall et A. Trivelpiece, Principles of Plasma Physis. Mc Graw Hill

[10] J.L. Lions, Quelques méthodes de resolution de problèmes aux limites linéaires Dunod.

[11] S. Ukai et T. Okabe, On the classical solution in the large in time of the two dimensional Vlasov Equation. Osaka J. of Math. n°15 (1978) 245-261.

[12] S. Wollman, On the Vlasov-Poisson and Vlasov Maxwell equations. Courant Institute Technical Report.

[13] L. Tartar. M.R.C. Tech. Rep. Univ. of Wis (1980)

[14] T. Nishida, Fluid dynamical limit of the non linear Boltzmann equation to the level of the compressible Euler equat Math. Phys. 61 119-148 (1978).

[15] P. Degond, à paraître.

Claude BARDOS
Université Paris-Nord
Département de Mathématique
Avenue J.B. Clément
93430 - VILLETANEUSE
FRANCE

Pierre DEGOND
Ecole Polytechnique
Centre de Mathématiques Appliquée
91128 - PALAISEAU CEDEX
FRANCE

TH. CAZENAVE & A HARAUX

On the nature of free oscillations associated with some semilinear wave equations

1. INTRODUCTION.

Dans toute la suite, Ω désigne un ouvert borné de $\mathbb{R}^n$, supposé connexe et $g : \mathbb{R} \times \Omega \times \mathbb{R} \to \mathbb{R}$ une fonction numérique des trois variables (t,x,u).

On s'intéresse au problème du comportement qualitatif lorsque $t \to \pm \infty$ des solutions du problème

$$\begin{cases} u \in C(\mathbb{R}, H_0^1(\Omega)) \cap C^1(\mathbb{R}, L^2(\Omega)) \\ u_{tt} - \Delta u + g(t,x,u(t,x)) = 0 \text{ sur } \mathbb{R} \times \Omega. \end{cases} \tag{1.1}$$

Dans le paragraphe 4, lorsque n=1, on se préoccupera aussi du cas où les conditions aux limites de Dirichlet sont remplacées par une condition de périodicité par rapport à x.

Divers cas particuliers de ce problème ont été déjà considérés dans la littérature. Plus précisément

- Le cas où g=0 a fait l'objet d'une étude intensive depuis d'Alembert [1]. Depuis 1970 on sait [2] que toutes les solutions de (1.1) sont presque-périodiques dans l'espace $H_0^1(\Omega) \times L^2(\Omega)$. Cette propriété a été démontrée successivement par plusieurs auteurs sous des hypothèses de plus en plus générales. La dernière démonstration, la plus simple et la plus transparente, est donnée par M. Biroli dans [5]. D'autre part, on sait depuis les travaux de Y. Meyer [11] et J.P. Allouche [3] que cette équation est en partie insensible aux propriétés de régularité des conditions initiales.

Plus précisément, la presque-périodicité de la fonction $t \to u(t,x)$ dans $C(\bar{\Omega})$ n'est obtenue que si $u(0)$ et $u_t(0)$ sont dans le domaine d'une puissance convenable de $(-\Delta)$ de telle sorte que cette propriété soit conservée avec le temps. En particulier, si n=2 et $\Omega =]0,1[\times]0,1[$, il existe une solution de (1.1) dans $C^1([0,T] \times \bar{\Omega})$ qui explose dans $C(\bar{\Omega})$ lorsque $t \to T$: cette propriété est évidemment liée au fait que $H^1(\Omega)$ n'est pas inclus dans $L^\infty(\Omega)$, et que le bornage de u(t,x) dans $H_0^1(\Omega)$ est le seul qui soit "retenu" par le système dynamique engendré par (1.1) dans l'espace des phases $H_0^1(\Omega) \times L^2(\Omega)$.

- Le cas où $g(t,x,u) = c|u|^{\alpha-1}u$, $\alpha > 1$ a été étudié par P. Rabinowitz [12] et par H. Brézis, J.M. Coron et L. Nirenberg [6]. Ces auteurs démontrent l'existence d'une solution $u \not\equiv 0$ et périodique en t de période 2ℓ lorsque $\Omega =]0,\ell[$.

Lorsque $\tau/\ell \notin Q$, on ne sait pas si (1.1) a des solutions périodiques de période τ en t. On sait encore moins décrire le comportement de la solution générale lorsque $t \to +\infty$.

Ce travail est motivé par l'idée très simple que sous des conditions qui empèchent l'existence de solutions stationnaires non nulles, toutes les solutions de (1.1) doivent représenter des "oscillations autour de 0".

Ceci est assez intuitif au moins lorsque $g(t,x,u) = g(u)$. Les résultats que nous énoncerons supposent toujours au moins $g(t,x,0) = 0$. Bien que partiels, nous espérons qu'ils aideront à éclaircir un jour la situation générale.

2. LE CAS OU g=0.

Dans ce paragraphe, on considère donc le problème

$$\begin{cases} u \in C(\mathbb{R},H_0^1(\Omega)) \cap C^1(\mathbb{R},L^2(\Omega)) \\ \\ u_{tt} - \Delta u = 0 \text{ dans } \mathcal{D}'(\mathbb{R} \times \Omega). \end{cases} \tag{2.1}$$

Pour toute solution de (2.1), la fonction $t \to u(t,x) \in H_0^1(\Omega)$ est fortement presque-périodique. Soit $0 < \lambda_1 < \lambda_2 \dots < \lambda_j < \dots$ la suite des valeurs propres de $(-\Delta)$ dans $H_0^1(\Omega)$. Le développement de u(t,x) en série de Fourier anharmonique est donné par

$$u(t,x) = \sum_{j=1}^{+\infty} \cos(\sqrt{\lambda}_j t+\alpha_j)u_j(x) \tag{2.2}$$

où l'on a posé

$$u_j(x) = \lim_{t\to+\infty} \frac{1}{t}\int_{a-t}^{a+t} u(s,x)\cos(\sqrt{\lambda}_j s+\alpha_j)ds \tag{2.3}$$

et la convergence dans (2.3) a lieu dans $H_0^1(\Omega)$ uniformément pour $a \in \mathbb{R}$.

Définissons un opérateur L auto-adjoint positif dans $L^2(\Omega)$ par

$$D(L) = H^2(\Omega) \cap H_0^1(\Omega),\ Lu = -\Delta u \text{ pour } u \in D(L).$$

Si on suppose qu'une solution u de (2.1) vérifie les deux conditions

$$\begin{cases} u(0,x) = u_0(x) \in D(L^{\frac{n}{4}+\varepsilon}) \\ u_t(0,x) = v_0(x) \in D(L^{\frac{n}{4}-\frac{1}{2}+\varepsilon}),\ \varepsilon > 0 \end{cases} \tag{2.4}$$

alors on a

$$u(t,x) \in C_B(\mathbb{R},H^{\frac{n}{2}+2\varepsilon}(\Omega)). \tag{2.5}$$

En particulier, dans ce cas pour tout $x \in \Omega$ la fonction $t \to u(t,x)$ est presque-périodique et son développement de Fourier anharmonique est donné par (2.2) *en tout point*. La théorie générale des fonctions presque-périodiques nous donne aussi que l'on a (cf [2])

$$\forall x \in \Omega,\quad \lim_{t\to+\infty} \left(\frac{1}{t}\int_0^t u(s,x)ds\right) = 0. \tag{2.6}$$

<u>Proposition 2.1</u>. *Soit* u *une solution de* (2.1) *vérifiant* (2.5). *Alors pour tout* $x_0 \in \bar{\Omega}$ *on a soit* $u(t,x_0)$ *identiquement nul, soit* $u(t,x_0)$ *prend des valeurs* > 0 *et des valeurs* < 0 *sur toute demi-droite de* $\mathbb{R}$. *De plus si* $u \not\equiv 0$, *la deuxième éventualité est réalisée pour presque tout point* $x_0 \in \Omega$.

Démonstration. Montrons d'abord que si $u(t,x_0) \geq 0$ pour $t \geq 0$, alors $u(t,x_0) = 0$ pour $t \in \mathbb{R}$.

Supposons par l'absurde que l'on ait

$$\begin{cases} \forall t \geq 0, \quad u(t,x_0) \geq 0 \\ \exists\, t_0 \in \mathbb{R} \text{ tel que } |u(t_0,x_0)| = 3\rho > 0. \end{cases} \tag{2.7}$$

La presque-périodicité de u implique l'existence de $\ell > 0$ et d'une suite de nombres $a_n \in \,]3n\ell,(3n+1)\ell[$ tels que

$$\sup_{t \in \mathbb{R}} |u(a_n+t,x_0) - u(t_0+t,x_0)| \leq \rho, \quad \forall n \in \mathbb{N} \tag{2.8}$$

En faisant t=0 dans (2.8) on voit que si $u(t_0,x_0) < 0$, alors $u(a_n,x_0) \leq -2\rho$, ce qui contredit (2.7). On a donc $u(t_0,x_0) = 3\rho$. Soit $\delta > 0$ tel que $\delta < \ell$ et $u(t_0+\sigma,x_0) \geq 2\rho$ lorsque $|\sigma| \leq \delta$. On a d'après (2.8) :

$$\forall n \in \mathbb{N}, \quad \forall \sigma \in [-\delta,\delta], \quad u(a_n+\sigma,x_0) \geq \rho. \tag{2.9}$$

En utilisant (2.9), on obtient

$$\limsup_{t \to +\infty} \left(\frac{1}{t}\int_0^t u(s,x_0)ds\right) \geq \frac{2\delta}{3\ell}\rho > 0. \tag{2.10}$$

Il est clair que (2.10) est contradictoire avec (2.6). Donc (2.7) est absurde et on a l'alternative annoncée.

Supposons maintenant que $u(t,x_0) = 0$ pour tout $t \in \mathbb{R}$. Du fait que (2.3) est vérifié en tout point donc en x_0, on déduit immédiatement que l'on a

$$\forall j \in \mathbb{N} \setminus \{0\}, \quad u_j(x_0) = 0 \tag{2.11}$$

Pour tout $j \in \mathbb{N} \setminus \{0\}$, $u_j(x)$ est une fonction propre de $L = (-\Delta)$ et cette fonction est donc analytique sur tout segment contenu dans Ω.

En particulier, si $u \not\equiv 0$, il existe $k \in \mathbb{N}$ tel que $u_k \not\equiv 0$ et l'ensemble des x_o qui vérifie (2.11) est nécessairement de mesure nulle.

Remarque 2.2. a) Si $x_o \in \Omega$ est tel que $\phi_j(x_o) = 0$ avec $L\phi_j = \lambda_j\phi_j$, alors $\cos(\sqrt{\lambda_j}t)\phi_j(x)$ est une solution de (2.1) très régulière qui est nulle en x_o pour $t \in \mathbb{R}$.

En particulier, *si L a des valeurs propres multiples, tout* point x_o est tel qu'il existe une solution non nulle u de (2.1) avec $u(t,x_o) \equiv 0$.

b) Si toutes les valeurs propres λ_j sont simples, et si $\{\phi_j(x)\}$ est une suite de fonctions propres normalisées, la réunion des zéros des ϕ_j dans $\bar{\Omega}$ est *un ensemble de mesure nulle* dans Ω en - dehors duquel toutes les solutions non nulles de (2.1), suffisamment régulières, doivent osciller indéfiniment.

c) Dans le cas n=1, $\Omega =]0,1[$, les zéros des fonctions propres sont exactement les points rationnels. Ici, toute solution de (2.1) étant périodique en t de période 2 et de moyenne nulle en t pour tout $x \in \Omega$, il est clair que si $u(t,x_o) \not\equiv 0$, la fonction $u(t,x_o)$ prend des valeurs > 0 et des valeurs < 0 sur tout intervalle J tel que $|J| \geq 2$.

En fait, dès que $|J| < 2$, on peut trouver des solutions de (2.1) non nulles qui restent ≥ 0 sur J pour tout $x_o \in U$ avec U ouvert $\subset \Omega$.

Plus précisément, soit $\phi \in C^\infty(\mathbb{R})$, périodique de période 2 et telle que $\phi(y) = 0$ sur $[\varepsilon,2]$ avec par exemple $\phi > 0$ sur $]0,\varepsilon[$.

La fonction

$$u(t,x) = \phi(t+x) - \phi(t-x)$$

est solution de (2.1).

Pour tout $x_o \in \Omega$, on a

$$\forall t \in [x_o+\varepsilon, x_o+2], \quad u(t,x_o) = \phi(t+x_o) \geq 0.$$

d) Pour résumer, en chaque point $x_o \in \Omega$ tel que $u(t,x_o) \not\equiv 0$, la fonction $t \to u(t,x_0)$ change de signe sur tout intervalle de longueur 2, et ce résultat est optimal pour la solution générale.

Le but de ce travail est d'obtenir des résultats du même type.

- En dimension 1 pour des problèmes semi-linéaires
- En dimension > 1, au moins dans le cas linéaire. [avec des résultats partiels pour les équations semi-linéaires]

3. LE CAS $\Omega =]0,1[$.

L'objet de ce paragraphe est la démonstration détaillée du résultat suivant annoncé dans [8]. Dans la suite, J désigne un intervalle fermé de $\mathbb{R}$ et $\mathring{J}$ son intérieur.

Théorème 3.1. *Soit* $g(t,u)$ *une fonction continue sur* $\mathbb{R}^2$*, continûment différentiable par rapport à* u*, impaire et croissante au sens large par rapport à* u *pour tout* $t \in \mathbb{R}$ *fixé. Soit* $\Omega =]0,1[$ *et* u *solution de*

$$\begin{cases} u \in C(J,H_0^1(\Omega)) \cap C^1(J,L^2(\Omega)) \\ u_{tt} - u_{xx} + g(t,u) = 0 \text{ dans } \mathcal{D}'(\mathring{J} \times \Omega) \end{cases} \qquad (3.1)$$

où J *est un intervalle [fermé] de* $\mathbb{R}$ *tel que* $|J| \geq 2$*. Supposons* $u \not\equiv 0$ *sur* $J \times \Omega$ *et posons*

$$d = \inf\{x \in \Omega,\ x > 0,\ u(t,x) = 0, \quad \forall t \in J\}$$

Alors on a les propriétés suivantes

1°) Il existe k *entier positif tel que* $d = \frac{1}{k}$.

2°) Pour tout m *entier compris entre* 0 *et* k*, on a*

$$\forall t \in J, \quad u(t,\tfrac{m}{k}) = 0$$

3°) Si x_0 *n'est pas de la forme* $\frac{m}{k}$ *avec* m *entier compris entre* 0 *et* k*,* *la fonction* $t \to u(t,x_o)$ *prend des valeurs* > 0 *et des valeurs* < 0 *sur tout intervalle fermé* $I \subset J$ *tel que* $|I| \geq \frac{2}{k}$.

Démonstration. Soit t_o un point intérieur à J et posons $u(t_o,x) = u_o(x)$, $u_t(t_o,x) = v_o(x)$. Soient respectivement $\tilde{u}_o(x)$ et $\tilde{v}_o(x)$ les prolongements impairs et 2-périodiques de u_o et v_o sur $\mathbb{R}$. Enfin, soit J' le plus grand intervalle ouvert $\subset$ J et contenant t_o sur lequel on peut résoudre le problème

$$\begin{cases} \tilde{u}_{tt} - \tilde{u}_{xx} + g(t,\tilde{u}) = 0 \text{ dans } \mathcal{D}'(J' \times \mathbb{R}) \\ \tilde{u}(t,x+2) = \tilde{u}(t,x) \text{ sur } J' \times \mathbb{R} \\ \tilde{u} \in C(J',H^1([0,2])) \cap C^1(J',L^2(0,2)) \\ \tilde{u}(t_o,x) = \tilde{u}_o(x),\ \tilde{u}_t(t_o,x) = \tilde{v}_o(x). \end{cases} \tag{3.2}$$

Comme g est impaire par rapport à u, il est immédiat de vérifier que

$$\tilde{u}(t,-x) = -\tilde{u}(t,x),\ \forall(t,x) \in J'\times \mathbb{R}. \tag{3.3}$$

En particulier on a $\tilde{u}(t,0) = \tilde{u}(t,1) = 0$, donc

$$\tilde{u}(t,x) = u(t,x),\quad \forall(t,x) \in J' \times \Omega. \tag{3.4}$$

Alors de (3.2), (3.3) et (3.4), puisque $J' \subset J$ on déduit grâce à (3.1) :

$$\tilde{u} \in L^\infty(J',H^1([0,2])) \cap W^{1,\infty}(J',L^2(0,2)). \tag{3.5}$$

Par définition de J' et grâce à l'existence locale pour (3.2), il est clair que (3.5) implique $\bar{J}' = J$. Donc $\tilde{u}$ est un prolongement de u sur $J \times \mathbb{R}$ qui vérifie

$$\begin{cases} \tilde{u}_{tt} - \tilde{u}_{xx} + g(t,\tilde{u}(t,x)) = 0 \text{ dans } \mathcal{D}'(\mathring{J} \times \mathbb{R}) \\ \tilde{u}(t,-x) = -\tilde{u}(t,x) \quad \text{sur } J \times \mathbb{R} \\ \tilde{u}(t,x+2) = \tilde{u}(t,x) \quad \text{sur } J \times \mathbb{R} \end{cases} \tag{3.6}$$

Les résultats du théorème 3.1 se déduiront aisément du lemme suivant.

Lemme 3.2. *Soit* $x_o \in \Omega$ *tel que* $u(t,x_o) \geq 0$ *pour tout* $t \in J$. *Si* $|J| \geq 2$, *on a*

$$\tilde{u}(t,x+2x_o) = \tilde{u}(t,x), \quad \forall (t,x) \in J \times \mathbb{R}. \tag{3.7}$$

Démonstration. On introduit la fonction

$$w(t,x) = \tilde{u}(t,x) + \tilde{u}(t,2x_o-x), \ \forall (t,x) \in J \times \mathbb{R}.$$

Grâce aux propriétés de g, on peut écrire

$$g(t,\tilde{u}(t,x)) + g(t,\tilde{u}(t,2x_o-x)) = h(t,x)w(t,x)$$

avec $h \in L^\infty(J \times \mathbb{R})$ et $h(t,x) \geq 0$.

La fonction w est alors solution de

$$\begin{cases} w_{xx} - w_{tt} = h(t,x)w(t,x) \text{ pour } (t,x) \in J \times \mathbb{R} \\ w(t,x_o) = 2u(t,x_o) \quad \forall t \in J \\ w_x(t,x_o) = 0 \quad \forall t \in J \end{cases} \tag{3.8}$$

Puisque $|J| \geq 2$, on peut supposer, après une translation sur la variable t, que $[0,2] \subset J$. Posons

$$\mathcal{T} = \{(t,x) \in J \times \mathbb{R}, \ |x-x_o| \leq t \leq 2 - |x-x_o|\}$$

On va maintenant établir le résultat suivant

$$\forall (t,x) \in \mathcal{T}, \quad w(t,x) \geq 0. \tag{3.9}$$

Il suffit pour cela de vérifier (3.9) lorsque $x \geq x_o$. Pour simplifier les notations, nous introduisons

$$\begin{cases} v(t,y) = w(t,x_0+y) \\ k(t,y) = h(t,x_0+y) \end{cases}$$

On a alors la formule élémentaire (cf. [10])

$$v(t,y) = u(t+y,x_0) + u(t-y,x_0) + \frac{1}{2}\int_0^y\int_{-(y-z)}^{y-z} k(t+\sigma,z)v(t+\sigma,z)d\sigma dz \qquad (3.10)$$

valable pour $0 \leq y \leq t \leq 2-y$.

Cette formule implique immédiatement

$$v^-(t,y) \leq \frac{1}{2}\int_0^y\int_{-(y-z)}^{y-z} k(t+\sigma,z)v^-(t+\sigma,z)d\sigma dz \qquad (3.11)$$

Soit avec $M = \frac{1}{2}\|h\|_{L^\infty(J \times \mathbb{R})}$

$$v^-(t,y) \leq M\int_0^y\int_{-(y-z)}^{y-z} v^-(t+\sigma,z)d\sigma dz$$

En posant $t+\sigma = \tau$, cette inégalité devient

$$v^-(t,y) \leq M\int_0^y\int_{t-(y-z)}^{t+y-z} v^-(\tau,z)d\tau dz \qquad (3.12)$$

Or : $z \leq t - y + z$ et $t + y - z \leq 2 - z$ pour tous (y,z) avec $0 \leq z \leq y \leq t \leq 2 - y$.

On déduit donc de (3.12) l'inégalité

$$v^-(t,y) \leq M\int_0^y\int_z^{2-z} v^-(\tau,z)d\tau dz. \qquad (3.13)$$

Finalement, soit $\psi(y) = \int_y^{2-y} v^-(\tau,y)d\tau$.

En intégrant (3.13) en t sur $[y,2-y]$ on trouve (puisque le membre de droite ne dépend pas de t) :

$$\psi(y) \leq 2M \int_0^y \psi(z)dz, \quad \forall y \in [0,1]. \tag{3.14}$$

Par hypothèse on a $\psi(0) = \int_0^2 v^-(\tau,0)d\tau$

$$= \int_0^2 w^-(\tau,x_0)d\tau = 2\int_0^2 \tilde{u}^-(\tau,x_0)d\tau = 0.$$

Il résulte de cette remarque et de (3.14) que $\psi(y) = 0$, $\forall y \in [0,1]$, d'où (3.9) lorsque $x \geq x_0$. Comme $w(t,x) = w(t,2x-x)$ on voit que (3.9) est vérifié aussi lorsque $x < x_0$.

De (3.9) on déduit en particulier en prenant t=1 :

$$\forall x \in [x_0-1,x_0+1], \quad \tilde{u}(1,x) + \tilde{u}(1,2x_0-x) \geq 0. \tag{3.15}$$

Grâce à l'imparité de $\tilde{u}$ on déduit de (3.15) que

$$\forall x \in [x_0-1,x_0+1], \ \tilde{u}(1,-x) \leq \tilde{u}(1,2x_0-x).$$

Soit en changeant x en (-x) :

$$\forall x \in [-x_0-1,-x_0+1], \quad \tilde{u}(1,2x_0+x) \geq \tilde{u}(1,x). \tag{3.16}$$

Puisque $\tilde{u}(1,x)$ est périodique de période 2, on déduit de (3.16) que l'on a en fait :

$$\forall x \in \mathbb{R}, \ \tilde{u}(1,2x_0+x) \geq \tilde{u}(1,x). \tag{3.17}$$

Soit $z(x) = \tilde{u}(1,2x_0+x) - \tilde{u}(1,x)$. Alors $z(x)$ est ≥ 0, périodique et de moyenne nulle : elle est donc nulle et on a

$$\forall x \in \mathbb{R}, \ \tilde{u}(1,2x_0+x) = \tilde{u}(1,x). \tag{3.18}$$

D'une façon rigoureusement identique, on peut vérifier que pour tout h tel que $0 < h < 1$, on a

$$\forall x \in [-x_0-1+h, -x_0+1-h], \quad \tilde{u}(1+h, 2x_0+x) \geq \tilde{u}(1+h, x). \tag{3.19}$$

En soustrayant (3.18) de (3.19), en divisant par h et en faisant tendre h vers 0, on trouve alors

$$\forall x \in]-x_0-1, x_0+1[, \quad \tilde{u}_t(1, 2x_0+x) \geq \tilde{u}_t(1, x). \tag{3.20}$$

Soit $\dot{z}(x) = \tilde{u}_t(1,2x_0+x) - \tilde{u}_t(1,x)$. Alors $\dot{z}$ est une fonction dans $L^2_{loc}(\mathbb{R})$, ≥ 0 presque partout et de moyenne nulle : elle est donc nulle presque partout, ce qui établit

$$\tilde{u}_t(1, 2x_0+x) = \tilde{u}_t(1, x) \quad \text{p.p. sur } \mathbb{R}. \tag{3.21}$$

Enfin, en utilisant l'invariance de (3.6) par translation en x et l'unicité des solutions locales avec condition de périodicité, on déduit de (3.18) et (3.21) que

$$\tilde{u}(t, 2x_0+x) = \tilde{u}(t, x), \quad \forall (t,x) \in J \times \mathbb{R} \tag{3.22}$$

Fin de la démonstration du théorème 3.1. Les détails sont donnés dans [8]. Du lemme 3.2 on déduit très facilement que $d > 0$ et $d \in Q$. En utilisant le théorème de Bezout et les propriétés de u on en déduit facilement 1°). Le point 2°) est une conséquence immédiate. Enfin, pour la démonstration du point 3°), on se place dans $\Omega' =]\frac{m}{k}, \frac{m+1}{k}[$ et si $x_0 \in \Omega'$ met en défaut la conclusion du 3°), on montre que $u(t, x_0 - \frac{m}{k}) \equiv 0$ pour $t \in J$ en utilisant le lemme 3.2. Cette contradiction avec la définition de d achève la démonstration du théorème 3.1.

Remarque 3.3. a) Au cours de la démonstration du théorème 3.1, on obtient des informations très précises sur la *structure* de l'ensemble des points x_0 tels que $u(t,x_0) \equiv 0$ et sur les conditions qui permettent à un point x_0 donné d'être dans cet ensemble. Voir à ce sujet les remarques de [8].

b) Dans le cas où g ne dépend pas de t, les solutions de (3.1) sont globales ($J=\mathbb{R}$).

En effet, en posant $G(u) = \int_0^u g(v)dv$, on a pour toute solution u de (3.1) la propriété de conservation :

$$\int_\Omega \{\tfrac{1}{2}u_x^2+\tfrac{1}{2}u_t^2 + G(u(t,x))\}dx = \text{constante pour } t \in J. \qquad (3.23)$$

Comme $G(u) \geq 0$ pour $u \in \mathbb{R}$, ceci implique l'existence globale. Dans ce cas, le théorème 3.1 permet d'obtenir des informations sur le comportement de $u(t,x)$ lorsque $t \to \pm\infty$. On a en effet le résultat suivant.

Corollaire 3.4. Soit g comme ci-dessus et u solution de (3.1). Alors on a

i) Soit $\lim\limits_{t\to+\infty} \operatorname{Sup}\limits_{x\in\Omega} |u(t,x)| = 0$

ii) Soit pour tout $x_0 \in \Omega \cap (\mathbb{R} \setminus Q)$, les inégalités

$$\liminf_{t\to+\infty} u(t,x_0) < 0 < \limsup_{t\to+\infty} u(t,x_0). \qquad (3.24)$$

Démonstration. De (3.23) on tire immédiatement que $\bigcup\limits_{t\in\mathbb{R}} \{u(t,.)\}$ est précompact dans $C(\bar{\Omega})$.

Supposons maintenant que ii) n'est pas vérifié avec par exemple $\limsup\limits_{t\to+\infty} u(t,x_0) \leq 0$, $x_0 \in \Omega \cap (\mathbb{R}\setminus Q)$.

Pour toute suite $\{t_n\}$ de réels tendant vers $+\infty$, on peut extraire une sous-suite, notée encore $\{t_n\}$ telle que

$$u(t_n+t,x) \to v(t,x)$$

dans $C(\mathbb{R},L^2(\Omega)) \cap L^2_{loc}(\mathbb{R},C(\bar{\Omega}))$.

Il est clair que v est une solution de (3.1) car $g(u(t_n+t,x)) \to g(v(t,x))$ dans $L^1_{loc}(\mathbb{R} \times \Omega)$ par exemple.

Or $v(t,x_0) \le 0$ pour tout $t \in \mathbb{R}$.

Grâce au théorème 3.1, 3°) on en déduit que $v \equiv 0$ sur $\mathbb{R} \times \bar{\Omega}$.

Comme la suite t_n était arbitraire, on en conclut que $u(t,x) \to 0$ dans $C(\bar{\Omega})$ lorsque $t \to +\infty$.

Ceci achève la démonstration du corollaire 3.4.

Remarques 3.5. a) Il est certain que l'on n'est pas toujours dans le cas i) : les solutions périodiques non triviales construites dans [6], [12] sont un exemple de cas où ii) est vérifié (mais dans ce cas 3.24 est une conséquence *triviale* du théorème 3.1). *On ne sait pas* si l'éventualité i) peut effectivement être réalisée.

b) Il serait également intéressant de savoir si, sans hypothèse d'imparité sur g, on peut encore montrer que $u(t,x_0)$ est soit identiquement nul, soit forcé de prendre des valeurs > 0 et des valeurs < 0 sur tout intervalle J tel que $|J| \ge 2$.

4. GENERALISATIONS DIVERSES.

Le résultat principal du §3 est que lorsque $\Omega =]0,1[$ et $x_0 \in \Omega$, toute solution $u \not\equiv 0$ de (3.1) vérifie la propriété suivante : ou bien $u(t,x_0) \equiv 0$, ou bien $u(t,x_0)$ change de signe sur tout intervalle de longueur ≥ 2.

Dans ce paragraphe, nous indiquons quelques résultats qui vont dans la même direction sans être aussi précis.

a) Un résultat d'oscillation global en espace.

Soit Ω comme dans l'introduction (avec $n \ge 1$ quelconque) et supposons par simplifier

$$g \in C^1(\mathbb{R} \times \bar{\Omega} \times \mathbb{R}) \tag{4.1}$$

Il existe $\gamma > 1$ tel que $(n-2)\gamma \le 4$ et $|g_u(t,x,u)| \le C(1+|u|^\gamma)$ sur $\mathbb{R} \times \bar{\Omega} \times \mathbb{R}$. (4.2)

On remplace l'hypothèse de monotonie par rapport à u par la condition plus faible

$$g(t,x,u)u \geq 0, \quad \forall (t,x,u) \in \mathbb{R} \times \Omega \times \mathbb{R}. \tag{4.3}$$

On a alors le résultat suivant.

Théorème 4.1. *Soit* g,Ω *comme ci-dessus et* $\lambda_1 > 0$ *la première valeur propre de* $(-\Delta)$ *dans* $H_0^1(\Omega)$. *Soit* u *une solution de* (1.1) *et supposons que l'on ait*

$$u(t,x) \geq 0 \quad \textit{p.p sur } J \times \Omega \tag{4.4}$$

$$|J| > \pi/\sqrt{\lambda_1}. \tag{4.5}$$

Alors en fait $u \equiv 0$ *p.p. sur* $\mathbb{R} \times \Omega$.

Démonstration. L'idée est de multiplier (1.1) par $\phi(x)$ et d'intégrer sur Ω où $\phi > 0$ est une fonction propre associée à λ_1. On pose

$$v(t) = \int_\Omega u(t,x)\phi(x)dx, \quad \forall t \in \mathbb{R}.$$

Alors $v \in C^2(\mathbb{R})$ et si $u \geq 0$ sur $J \times \Omega$ on obtient

$$v \geq 0 \text{ et } v'' \leq -\lambda_1 v(t), \quad \forall t \in J. \tag{4.6}$$

Les inégalités (4.6) entraînent que si $|J| \leq \frac{\pi}{\sqrt{\lambda_1}}$, soit $v \equiv 0$ sur J. Dans le deuxième cas on a $u \equiv 0$. Pour les détails nous renvoyons à [9].

Remarques 4.2. 1) Le résultat du théorème 4.1 est optimal lorsque $g(t,x,u) = o(|u|)$ uniformément par rapport à t et x. En particulier lorsque $g = 0$ la fonction $\sin(\sqrt{\lambda_1}t)\phi(x)$ est solution de (1.1) et est ≥ 0 sur $[0,\frac{\pi}{\sqrt{\lambda_1}}] \times \Omega$.

2) Lorsque $n = 1$ et $\Omega =]0,1[$, on sait que $u(t,x_0)$ peut rester ≥ 0 sur $[0,2-\varepsilon]$ avec ε arbitrairement petit et x_0 bien choisi. Sur l'exemple de la remarque 2.2, c) , on voit que "le plus long intervalle où $u(t,x_0) \geq 0$", bien que de longueur toujours > 1, dépend de x_0. Ainsi la conclusion du théorème 4.1 (qui dit que u ne peut être ≥ 0 sur $J \times \Omega$ avec $|J| > 1$) n'est

pas contredite, et le "temps de positivité locale" est en général supérieur au "temps de positivité globale".

b) Autres conditions aux limites homogènes sur $\Omega =]0,1[$.

Le cas des conditions de Neumann s'y ramenant par prolongement, nous nous limiterons au cas du problème périodique

$$\begin{cases} u \in C(\mathbb{R},H^1(\Omega)) \cap C^1(\mathbb{R},L^2(\Omega)) \\ u_{tt} - u_{xx} + g(t,u) = 0 \text{ sur } \mathbb{R} \times \mathbb{R} \\ u(t,x+1) = u(t,x) \text{ sur } \mathbb{R} \times \mathbb{R}. \end{cases} \tag{4.7}$$

Dans le cas où $g = 0$, toute constante réelle est solution de (4.7) : on ne peut donc espérer obtenir le caractère oscillant de *toutes* les solutions que si g vérifie une condition de coercivité

Théorème 4.3. *Supposons que l'on ait* $g \in C^1(\mathbb{R}^2)$ *et*

$$g(t,u) \geq f(t)u^p, \quad \forall (t,u) \in \mathbb{R}^+ \times \mathbb{R}^+ \tag{4.8}$$

où $p \geq 1$ *et* $f \geq 0$ *est telle que*

$$\int_0^{+\infty} f(s)ds = +\infty. \tag{4.9}$$

Alors si une solution u *de* (4.7) *est telle que* $u(t,x) \geq 0$ *sur* $\mathbb{R}^+ \times \Omega$, *on a* $u \equiv 0$.

Démonstration. On introduit la fonction $v(t) = \int_\Omega u(t,x)dx$. En intégrant (4.7) sur Ω et en utilisant (4.8) on trouve que si $u(t,x) \geq 0$ sur $\mathbb{R}^+ \times \Omega$, alors $v \geq 0$ sur $\mathbb{R}^+$ et $v''(t) \leq - f(t) \int_\Omega (u(t,x))^p dx \leq - f(t)(v(t))^p$

Il est clair que $v(t)$ est alors croissante et tend vers une limite $\ell \geq 0$ (éventuellement infinie) tandis que $v'(t)$ est décroissante au sens large.

Si $\ell > 0$, il existe $t_0 \in \mathbb{R}^+$ tel que

$$v'' \leq -\delta f(t) \text{ pour } t \geq t_0, \quad \delta > 0. \tag{4.10}$$

De (4.10) on tire par intégration

$$v'(t) \leq v'(t_0) - \delta \int_{t_0}^{t} f(s)ds$$

Comme $\int_{t_0}^{+\infty} f(s)ds = +\infty$, il existe donc $t_1 \geq t_0$ tel que $v'(t_1) < 0$. On a alors une contradiction. Il en résulte automatiquement que $\ell = 0$, d'où $v(t) \equiv 0$ pour $t \geq 0$. On en conclut que $u(t,x) \equiv 0$ pour $(t,x) \in \mathbb{R}^+ \times \Omega$, puis $u \equiv 0$.

Remarque 4.4. La fonction $u(t,x) = \sqrt{t}$ est solution de $u_{tt} = -\frac{1}{4}\frac{1}{t^2}u(t,x)$ pour $t \geq 1$. Il est facile de la prolonger par une solution C^∞ de (4.7) avec $g(t,u) = \frac{1}{4t^2}u$ pour $t \geq 1$ et $g \in C^\infty(\mathbb{R} \times \mathbb{R})$. Cet exemple montre que la conclusion du théorème 4.3 peut être violée si (4.9) n'est pas vérifiée.

On a aussi le résultat suivant analogue au théorème 3.1.

Théorème 4.5. *Soit* $g \in C^1(\mathbb{R}^2)$ *impaire et croissante au sens large en* u *à* t *fixé, avec*

$$(g(t,u)-g(t,v))(u-v) \geq f(t)|u-v|^{p+1}, \quad \forall (t,u,v) \in \mathbb{R}^3 \tag{4.11}$$

où $p \geq 1$ *et* $f \geq 0$ *vérifie* (4.9).

Alors si u *est une solution quelconque de* (4.7), *pour tout* $x_0 \in \Omega$, *on a*

- *soit* $u(t,x_0) \equiv 0, \quad \forall t \in \mathbb{R}$

- *soit la fonction* $t \to u(t,x_0)$ *change de signe indéfiniment lorsque* $t \to +\infty$.

Démonstration. On introduit $w(t,x) = u(t,x) + u(t,2x_0-x)$ qui est périodique de période 1 en x et vérifie $w_{tt} - w_{xx} + h(t,x)w(t,x) = 0$ avec $h \geq 0$. Si on suppose par exemple $u(t,x_0) \geq 0$ pour $t \geq 0$, alors $w \geq 0$ sur $[1,+\infty[\times \Omega$.

Ensuite en utilisant (4.11), on établit par la méthode de démonstration du théorème 4.3 que $w \equiv 0$ sur $\mathbb{R} \times \Omega$.

On en conclut en particulier que $u(t,x_0) \equiv 0$.

c) Solutions radiales dans une boule.

Soit $\Omega = B(0,1)$ la boule de centre 0 et de rayon 1 dans $\mathbb{R}^n$, $n > 1$.

Soit u une solution de (1.1) suffisamment régulière et *radiale* (on suppose $g(t,x,u) = g(t,|x|,u))$. On peut écrire (1.1) sous la forme

$$u_{rr} + \frac{n-1}{r} u_r = u_{tt} + g(t,r,u). \qquad (4.12)$$

Si g est suffisamment régulière et vérifie (4.3), on peut établir le résultat suivant.

Proposition 4.6. *Pour toute solution* $u \not\equiv 0$ *comme ci-dessus, la fonction* $t \to u_r(t,1)$ *prend des valeurs* > 0 *et des valeurs* < 0 *sur tout intervalle tel que*

$$|J| \geq 2$$

En utilisant la théorie des fonctions presque-périodiques, on peut en déduire le résultat suivant.

Corollaire 4.7. *Soit* $\lambda \in C^\infty([0,1])$, $\lambda(r) \geq 0$ *et soit* $u \in C^\infty(\mathbb{R} \times \bar{\Omega})$ *une solution radiale* $\not\equiv 0$ *de*

$$\begin{cases} u_{tt} - \Delta u + \lambda(r)u = 0 & \text{sur } \mathbb{R} \times \Omega \\ u(t,x) = 0 & \text{sur } \mathbb{R} \times \partial\Omega \end{cases} \qquad (4.13)$$

Alors il existe un ensemble F fini, $F \subset]0,1[$ tel que pour tout $\rho \in [0,1[\setminus F$, la fonction $t \to u(t,x_0)$ prend des valeurs > 0 et des valeurs < 0 sur tout intervalle J de longueur ≥ 2 , pour tout $x_0 \in \Omega$ tel que $|x_0| = \rho$.

Remarque 4.8. L'ensemble F dépend de u mais il est contenu dans une partie dénombrable fixe de $]0,1[$, à savoir la réunion des zéros des fonctions propres de l'opérateur $u = - u_{rr} - \frac{n-1}{r} u_r + \lambda(r)u$ dans $H^1_{0,r}(\Omega)$.

d) Corde vibrante avec obstacle rectiligne.

Soit $\Omega =]-\frac{1}{2},\frac{1}{2}[$ et $h > 0$. On s'intéresse aux "solutions conservant l'énergie" du problème

$$u \in C(\mathbb{R},H^1_0(\Omega)) \cap W^{1,\infty}(\mathbb{R},L^2(\Omega)),\ u_{tt} - u_{xx} \in - g(u) \qquad (4.14)$$

$$\text{avec } g(u) = \begin{cases} 0 & \text{si } u > - h \\ ]-\infty,0] & \text{si } u = - h \\ \phi & \text{si } u < - h \end{cases}$$

La bonne formulation du Problème de Cauchy associé à (4.14) a été introduite dans [14].

- Il est clair que si $u(t,x) \geq 0$ sur $J \times \Omega$ avec $|J| > 1$, alors $u \equiv 0$. En effet, dans ce cas u est solution de $u_{tt} - u_{xx} = 0$ dans $\mathcal{D}'(\mathring{J} \times \Omega)$ et en appliquant le théorème 4.1 avec $g = 0$ on obtient $u = 0$ p.p. sur $J \times \Omega$, d'où $u \equiv 0$ par unicité du Problème de Cauchy mentionné ci-dessus.

- Lorsque $u(0,x) = u_0(x)$ est une donnée initiale "unimodale" et $u_t(0,x) = 0$, on a établi dans [7] que $u(t,x)$ est une fonction presque-périodique du temps, *qui en général n'est pas périodique*.

D'autre part, même si $u_0 \in C^\infty(\bar{\Omega})$, en général la dérivée $u_t(t,x)$ *n'est pas continue* de $\mathbb{R}$ dans $\mathcal{D}'(\Omega)$: ceci est dû aux chocs contre l'obstacle rectiligne d'équation $u = - h$.

Bien que l'évolution du système soit assez difficile à suivre dans la région où les chocs apparaissent, on a l'information *à priori* suivante.

Proposition 4.9. *Supposons que* $u(0,x) \in H^2(\Omega) \cap H_0^1(\Omega)$ *et* $u_t(0,x) \in H_0^1(\Omega)$. *Alors si* $u(t,x) \le 0$ *sur* $J \times \Omega$ *avec* $|J| > 1$, *on a* $u \equiv 0$.

Démonstration. Pour tout $\varepsilon > 0$ on considère la solution $u^\varepsilon(t,x)$ de

$$\begin{cases} u^\varepsilon \in C(\mathbb{R},H_0^1(\Omega)) \cap C^1(\mathbb{R},L^2(\Omega)) \\ \\ u_{tt}^\varepsilon - u_{xx}^\varepsilon = \frac{1}{\varepsilon}(u_\varepsilon+h)^- \text{ sur } \mathbb{R} \times \Omega \end{cases} \tag{4.15}$$

D'après [4], on sait que $u^\varepsilon \to u$ dans $C(a,b;L^2(\Omega))$ pour tous $(a,b) \in \mathbb{R}^2$, $a < b$.

Soit $v(t) = \int_{-1/2}^{1/2} u(t,x)\cos(\Pi x)dx$.

Si $v(t) < 0$ pour $t \in J$, où J est un intervalle fermé avec $|J| > 1$, alors pour $\varepsilon > 0$ assez petit on a

$$v_\varepsilon(t) = \int_{-1/2}^{1/2} u^\varepsilon(t,x)\cos(\Pi x)dx < 0, \quad \forall t \in J.$$

En multipliant (4.15) par $\cos \Pi x$ et en intégrant sur Ω, on trouve d'autre part $v_\varepsilon \in C^2(J)$ et $v_\varepsilon'' + \pi^2 v_\varepsilon \le 0$, $\quad \forall t \in J$.

Ceci est impossible si $|J| > 1$.

Il existe donc $t \in J$ tel que $v(t) \ge 0$. En fait, on peut même trouver $\delta > 0$ tel qu'il existe t_0 avec $[t_0-\delta,t_0+\delta] \subset J$ et tel que $v(t_0) \ge 0$. Comme $u(t_0,x) \le 0$ pour $x \in \Omega$ on en déduit que

$$u(t_0,x) = 0, \quad \forall x \in \Omega. \tag{4.16}$$

Enfin on déduit de (4.16) que $u \in C^1([t_0-\eta,t_0+\eta];L^2(\Omega))$ avec $\eta > 0$ et comme $u(t_0+h,x) \le 0$ pour $|h| < \delta$ on obtient

$$u(t_0+h,x) - u(t_0,x) \le 0 \text{ pour } |h| < \delta,\ x \in \Omega. \tag{4.17}$$

En divisant (4.17) par h et faisant $h \to 0$, on trouve finalement

$$u_t(t_o,x) = 0 \text{ p.p. sur } \Omega. \tag{4.18}$$

De (4.16) et (4.18) on conclut que $u \equiv 0$. Ceci achève la démonstration de la proposition 4.9.

Remarque 4.10. Ce résultat soulève deux questions naturelles

- Peut-on éliminer la condition $(u_o,v_o) \in H^2 \times H^1$?
- Que se passe-t-il du point de vue *local* ?

REFERENCES.

[1] J. d'Alembert, Opuscules mathématiques. David éditeur, Paris (1761).

[2] L. Amerio, G. Prouse, Abstract Almost Periodic Functions and Functional Equations. Van Nostrand, New-York (1971).

[3] J.P. Allouche, Thèse, Bordeaux (Mai 1983).

[4] A. Bamberger, M. Schatzman, New results on the vibrating string with a continuous obstacle, M.R.C. Technical Symmary Report, Madison, Wisconsin, # 2073 (1980).

[5] M. Biroli, Sur les solutions bornées et presque périodiques des équations et inéquations d'évolution. Ann. Mat. Pura Appl. 93, (1972) 1-79.

[6] H. Brézis, J.M. Coron, L. Nirenberg, Free vibrations for a nonlinear wave equation and a theorem of P. Rabinowitz. C.P.A.M. 33, (1980) 667-689.

[7] H. Cabannes, A. Haraux, Mouvements presque périodiques d'une corde vibrante en présence d'un obstacle rectiligne. C.R.A.S. Paris 191 A, (1980) 563-565.

[8] T. Cazenave, A. Haraux, Propriétés oscillatoires des solutions de certaines équations des ondes semi-linéaires. C.R.A.S. Paris Série A, à paraitre (1984).

[9] T. Cazenave, A. Haraux, Oscillatory behavior of the solutions to some semilinear wave equations, à paraître.

[10] R. Courant, D. Hilbert, Methods of Mathematical Physics. Vol. II, Interscience Publish. New-York, London (1962).

[11] Y. Meyer, Communication personnelle.

[12] P. Rabinowitz, Free vibrations for a semilinear wave equation, C.P.A.M 31, (1978) 31-68.

[13] C. Reder, Etude qualitative d'un problème hyperbolique avec contrainte unilatérale. Thèse de 3e cycle, Université de Bordeaux (1979).

[14] M. Schatzman, An hyperbolic problem of second order with unilateral constraints : the vibrating string with a concave obstacle. J. Math. Anal. Appl. 73, (1980) 138-191.

Thierry CAZENAVE & Alain HARAUX

Laboratoire Analyse Numérique
Tour 55-65, 5e étage
Université Pierre et Marie Curie
4, place Jussieu

75230 PARIS CEDEX 05

FRANCE

D DE FIGUEIRO

On the uniqueness of positive solutions of the Dirichlet problem $-\Delta u = \lambda \sin u$

1. INTRODUCTION.

Let Ω be a smooth bounded domain in R^N, $N \geq 2$. Consider the problem

$$-\Delta u = \lambda \sin u \text{ and } u > 0 \text{ in } \Omega, \ u = 0 \text{ on } \partial\Omega. \qquad (1)$$

Let λ_1 denote the first eigenvalue of $(-\Delta, H_0^1(\Omega))$, and $\phi_1 > 0$ a corresponding eigenfunction. It is well known that for $\lambda \leq \lambda_1$, problem (1) has no solution, and for $\lambda > \lambda_1$, problem (1) possesses a solution u, with $\|u\|_{L^\infty} < \pi$. Such a solution is obtained by the method of monotone iterations, using as subsolution $\underline{u} = \varepsilon\phi_1$, where ε is a sufficiently small positive number, and as supersolution $\bar{u} = \pi$. It follows then that (1) has a minimal solution u_{min} and that all eventual solutions u of (1), with $u(x) < \pi$, are such that $u_{min}(x) < u(x)$ for all $x \in \Omega$. Then by an argument of comparison of eigenvalues (see Berestycki [1]), problem (1) *with the additional requirement* that $\|u\|_{L^\infty} < \pi$, possesses exactly one solution, when $\lambda > \lambda_1$.

An interesting question is to ascertain whether or not problem (1) has solutions with $\|u\|_{L^\infty} > \pi$. In dimension N = 1, using ordinary differential equations arguments, namely the study of the problem in the phase plane, one sees that all solutions of (1) are bounded by π. In this note we establish that this is also the case in dimensions $N \geq 2$ for certain classes of domains Ω. Namely, *all solutions* u *of* (1) *are such that* $\|u\|_{L^\infty} < \pi$. This result, as observed above, implies that problem (1) has exactly one solution, when $\lambda > \lambda_1$ and Ω satisfies the Inheritance Assertion, see below.

This a priori bound on the solutions of (1) is obtained using Pohozaev's identity [2] and relies heavily on the Inheritance Assertion, stated next. First observe that by Sard's lemma there exists a set A of real numbers in the interval $[0, \|u\|_{L^\infty}]$ for each solution u of (1), with the properties that the Lebesgue measure of A is equal to $\|u\|_{L^\infty}$ (in particular, A is dense in the said interval) and that for each $a \in A$ the set $\Gamma_a = \{x \in \Omega : u(x) = a\}$ is a regular hyper-surface in R^N. These "level lines" or "level surfaces" a priori may be disconnected.

Inheritance Assertion. *For starshaped regions Ω the level lines Γ_a, with $a \in A$, are also starshaped.*

At present we do not know if the Inheritance Assertion is true for all starshaped regions. In the basis of the results of Gidas-Ni-Nirenberg [3] we see that there are regions Ω for which such an assertion holds true. For example, if Ω is a ball the solutions of (1) are radial symmetric and in fact Γ_a is a sphere for all a's. It has been proved by Kawohl [4] that if the region has some symmetry properties (ellipsoids or even some non convex regions) the above assertion holds for all a's.

Since the arguments below apply to other nonlinearities besides sin u, we state and prove a more general result, which implies readily the uniqueness of solution of problem (1).

2. A PRIORI BOUNDS VIA POHOZAEV'S IDENTITY.

Consider the problem

$$-\Delta u = f(u) \text{ and } u > 0 \text{ in } \Omega,\ u = 0 \text{ on } \partial\Omega, \tag{2}$$

where f satisfies the following condition

(f1) $\quad f : R^+ \to R$ is a $C^{N-2,\alpha}$-function.

Such a degree of regularity is only required to assert that the eventual H_0^1-solutions of (2) are C^N functions. So by Sard's lemma almost all level sets of the solutions u of (2) are regular hypersurfaces, just like the case of (1) discussed in the previous section. We shall use the same notations A and Γ_a for the corresponding sets in the case of (2). Next $F(s) = \int_0^s f$ and assume (f2) $F_\infty \equiv \sup\{F(s) : s > 0\} < \infty$ and $S \equiv \{s > 0 : F(s) = F_\infty\} \neq \emptyset$.

Let $M = \inf S$. Then the following result holds.

<u>Theorem</u>. *Let* f *satisfy conditions* (f1) *and* (f2). *Suppose that* Ω *is starshaped and that the Inheritance Assertion holds. Then* $\|u\|_{L^\infty} < M$ *for all solutions* u *of* (2).

<u>Remarks</u> 1) The theorem holds if Ω is a ball, an ellipsoid or even some non-convex regions with appropriate symmetries as described in [4]. This follows from the results of [3] which give in these cases the level lines of u inherit the geometry of $\partial\Omega$.

2) If the function f is such that f(s)/s is non-increasing for $0 < s < M$, one obtains immediately the uniqueness of solution of (2), using comparison of eigenvalues as in [1].

<u>Proof of the theorem</u>. (i) It follows by an application of the maximum principle (see for instance, Ambrosetti-Hess [5]) that $\|u\|_{L^\infty} \notin S$ for all solutions u of (2). ii) Next suppose that there exists a solution u of (2) with $c \equiv \|u\|_{L^\infty} > M$. By (i) above one has $F(c) < F_\infty$. Let $b = \max\{s < c : F(s) = F_\infty\}$. Then one can choose $a \in (b,c) \cap A$ such that $F(a) > F(s)$ for all $s \in [a,c]$. Now let $\Omega_a = \{x \in \Omega : u(x) > a\}$. Observe that Ω_a is starshaped and $\partial\Omega_a$ is a regular hypersurface. Define $v(x) = u(x) - a$ for $x \in \bar{\Omega}_a$. This function v is a solution of the problem

$$-\Delta v = g(v) \text{ and } v > 0 \text{ in } \Omega_a,\ v = 0 \text{ on } \partial\Omega_a, \tag{3}$$

where $g(s) = f(s+a)$. The solutions of (3) satisfy Pohozaev's identity

$$2N \int_{\Omega_a} G(v) - (N-2)\int_{\Omega_a} |\nabla v|^2 = \oint_{\partial\Omega_a} (x,\nu)|\nabla v|^2.$$

So the statement that (3) has a solution implies that $G(s) > 0$ for some s in the interval $I = [0, \|v\|_{L^\infty}]$. However, since $G(s) = F(s+a) - F(a)$ one comes to a contradiction. Q.E.D.

REFERENCES

[1] H. Berestycki, "Le nombre de solutions de certains problèmes semi-linéaires elliptiques", J. Fctl. Anal. 40 (1981), p. 1-29.

[2] S.I. Pohozaev - "Eigenfunctions of $\Delta u + \lambda f(u) = 0$". Soviet Math. Dokl 6 (1965) p. 1408-1411.

[3] B. Gidas, W-M. Ni and L. Nirenberg - "Symmetry and related properties via the maximum principle". Comm. Math. Phys. 68 (1979) pp. 209-243.

[4] B. Kawohl,

[5] A. Ambrosetti and P. Hess - "Positive solutions of asymptotically linear elliptic eigenvalue problems". J. Math. Anal. Appl. 73 (1980), p. 411-422.

Djairo G. DE FIGUEIREDO
Departamento de Matematica
Universidade de Brasilia
70910 BRASILIA
D.F. Brasil

L FRANK & F WENDT

Elliptic and parabolic singular perturbations in the kinetic theory of enzymes

INTRODUCTION.

In the Kinetic theory of membranes with enzymotic activity, a second order semilinear parabolic operator appears to be a suitable mathematical model for describing the dynamical process for the corresponding concentrations. The nonlinearity of this operator is affected by the presence of two positive parameters : ε, the so-called Michaelis' constant, and λ, the latter being connected with the ratio of initial concentrations of the enzyme and the substratum. For several realistic membranes, ε is small compared to λ and the data of the problem ([12]).

For each $\varepsilon > 0$ fixed, the mathematical model above fits the classical framework of Fréchet differentiable nonlinear operators. One can view this model as a family of perturbations (regular or singular, according to the magnitude of λ) of some reduced parabolic operator with a piecewise constant discontinuous nonlinearity. The same is also true for the corresponding elliptic stationary problem.

Further, for the stationary problem, there exists a critical value λ_c of the second parameter λ, such that for $\lambda < \lambda_c$ the original problem is a regular perturbation of the "reduced" one, whereas for $\lambda > \lambda_c$, it becomes a singular perturbation and is characterized by the presence of boundary layers located in a neighbourhood of the free boundary of the solution to the "reduced" problem. The set E_c of zeroes of the corresponding critical solution u^{λ_c} plays an important role in the investigation of the reduced problem.

One of the central results presented here is the sharp error estimate in the H_1-norm for the difference of the solutions to the perturbed and reduced problems. Further, λ_c is investigated as a functional of the data of the reduced problem.

Now the contents of the paper will be briefly sketched.

Part I deals with the stationary problem. Section 1.1 contains existence, uniqueness and regularity results as well as continuity results for the operators considered. In section 1.2, asymptotic solutions (for $\varepsilon \downarrow 0$) of the perturbed stationary problem are constructed under some regularity asumptions on the free boundary and the sharp H_1-estimate for the difference w between the solutions of the reduced and the perturbed problems is established. Namely, the following two-sided estimate holds with some constant C : $C^{-1}\ \varepsilon^{3/4} \leq \|w_\varepsilon\|_{H_1} \leq C\varepsilon^{3/4}$. The special case of a piecewise linear nonlinearity plays an important role and allows a considerable simplification in the construction of asymptotic solutions in the general case, as well. A one-sided estimate of the form $\|w_\varepsilon\|_{H_1} \leq C\varepsilon^{1/2}$, whose proof was merely based on the monotonicity of the nonlinearity and did not require regularity assumptions on the free boundary, had previously been given in [2], (see also [16]).

In section 1.3, properties of the critical value λ_c are stated and an improved convergence result of the form $\|w_\varepsilon\|_{H_1} \leq C\varepsilon(\lambda_c-\lambda)^{-1}$ is established when $\lambda < \lambda_c$. Sections 1.4 and 1.5 contain investigation of λ_c as a functional of the data.

Estimates of λ_c from above and from below are indicated and the formula for the Fréchet derivative of λ_c stated in [6,8] is proved. In section 1.6, the asymptotic behaviour for $\lambda \to \infty$ of the solution of the reduced problem and of the corresponding free boundary are investigated.

In Part II, the nonstationary problem is considered.

Section 2.1 contains existence, uniqueness and regularity results. In section 2.2, it is shown that the solution of the stationary problem is asymptotically stable in H_1 uniformly with respect to ε. In section 2.3, an estimate for the difference of the solutions to the nonstationary perturbed and reduced problems is proved. Sections 2.4 and 2.5 deal with nonnegative solutions and special solutions of the Cauchy problem for the reduced equation.

Several results in this paper have been announced in [6-8].

The list of references contains essentially papers which are (to the best of the authors' knowledge) tightly connected with the topics presented here. We refer to [15,14,4], [3], [5,11], [1,13,10] and [16,18-20] for more information concerning variational inequalities, maximal monotone operators, variational calculus and convex analysis, free boundary problems and singular perturbation theory.

0. NOTATION. STATEMENT OF THE PROBLEM.

Let $U \subset \mathbb{R}^n$ be a bounded domain with C^∞-boundary ∂U and let $\mathbb{R}_+ = (0,\infty)$. Denote,

$$Q = U \times \mathbb{R}_+,\ Q_T = U \times (0,T),\ \Gamma = \partial U \times \mathbb{R}_+,\ \ \Gamma_T = \partial U \times [0,T). \qquad (0.1)$$

Let the function $f \in C^0(\mathbb{R},\mathbb{R})$ be piecewise continuously differentiable : $f \in C^1([s_k,s_{k+1}])$ $\forall k \in \{0,\dots,r-1\}$, where $-\infty = s_0 < s_1 < \dots < s_r = \infty$. It is also assumed that

$$\begin{cases} f(0) = 0 \\ 0 \le f'(s) \le L(1+s^2)^{-1} \quad \forall s \in \mathbb{R} \setminus \{s_1,\dots,s_{r-1}\}, \end{cases} \qquad (0.2)$$

where $L > 0$ is constant.

As a consequence of (0.2), $f(s)$ is monotonically increasing on $\mathbb{R}$ and, moreover, there exist the limits

$$\lim_{s\to\pm\infty} f(s) = f_{\pm\infty}, \quad -\infty < f_{-\infty} \le f_{+\infty} < +\infty.$$

$\mathcal{H}(s)$ being Heaviside's function, we associate with $f(s)$ the following function

$$f_0(s) = f_{+\infty}\mathcal{H}(s) + f_{-\infty}\mathcal{H}(-s), \quad \forall s \in \mathbb{R} \setminus \{0\},\ \ f_o(0) = 0. \qquad (0.3)$$

We also denote by $F(s)$ and $F_0(s)$ the primitive functions of $f(s)$ and $f_0(s)$ normalized by the condition : $F(0) = 0$, $F_0(0) = 0$. Let $(a_{kj}(x,t))_{1\le k,j\le n}$ be uniformly with respect to $(x,t) \in \bar{Q}$ positive definite, and let $a_{kj}(x,t) \in C^{\infty}(\bar{Q})$. It is assumed that the family

$$t \to A(x,t,\frac{\partial}{\partial x}) = - \sum_{1\leq k,j\leq n} \frac{\partial}{\partial x_j} a_{kj}(x,t) \frac{\partial}{\partial x_k}, \quad a_{kj} \in C^{\infty}(\bar{Q}); \tag{0.4}$$

stabilizes, as $t \to +\infty$, to the operator :

$$A_{\infty}(x,\frac{\partial}{\partial x}) = - \sum_{1\leq k,j\leq n} \frac{\partial}{\partial x_j} a^{\infty}_{kj}(x) \frac{\partial}{\partial x_k}, \quad a^{\infty}_{kj} \in C^{\infty}(\bar{U}). \tag{0.5}$$

The following initial-boundary value problem $\mathfrak{T}^{\lambda}_{\varepsilon}$ is considered :

$$\begin{cases} \dfrac{\partial u^{\lambda}_{\varepsilon}}{\partial t} + A(x,t,\dfrac{\partial}{\partial x})u^{\lambda}_{\varepsilon} + \lambda f(\dfrac{u^{\lambda}_{\varepsilon}}{\varepsilon}) = g(x,t), & (x,t) \in Q \\ u^{\lambda}_{\varepsilon}(x,0) = \psi(x), & x \in \bar{U} \\ \pi_o u^{\lambda}_{\varepsilon}(x',t) = \phi(x',t), & (x',t) \in \Gamma \end{cases} \tag{0.6}$$

where λ,ε are positive parameters, π_0 is the restriction operator to Γ, the data is supposed to have the following regularity :

$$g \in C^{0}(\bar{Q}), \quad \psi \in C^{2}(\bar{U}), \quad \phi \in C^{2,1}(\Gamma),$$

to satisfy the compatibility condition :

$$\pi_0\psi(x') = \phi(x',0), \quad \forall x' \in \partial U \tag{0.7}$$

and to stabilize to

$$g_{\infty} \in C^{0}(\bar{U}), \quad \phi_{\infty} \in C^{2}(\partial U),$$

as $t \to +\infty$.

Here, as usual, $C^{0}(\bar{\Omega})$ is the space of all continuous in $\bar{Q}$ real-valued functions, $C^{2}(\bar{U})$ is the space of all twice continuously differentiable real-valued functions in $\bar{U}$ and $C^{2,1}(\Gamma)$ is the space of all continuous real-valued functions on Γ such that their first derivatives with respect to $(x',t) \in \Gamma$ and the second derivatives with respect to $x' \in \partial U$ are continuous functions. The parameter ε is assumed to be small compared to λ : $0 < \varepsilon << \lambda$.

Along with the problem $\mathfrak{T}_\varepsilon^\lambda$, we also consider the corresponding stationary problem $\mathfrak{T}_{\varepsilon,\infty}^\lambda$.

$$\begin{cases} A_\infty(x,\frac{\partial}{\partial x})u_{\varepsilon,\infty}^\lambda + \lambda f(\frac{u_{\varepsilon,\infty}^\lambda}{\varepsilon}) = g_\infty(x), & x \in U \\ \pi_0 u_{\varepsilon,\infty}^\lambda(x') = \phi_\infty(x') & x' \in \partial U. \end{cases} \tag{0.8}$$

Example 0.1. With $A = -\Delta$, $f(s) = s(1+|s|)^{-1}$, the problem (0.6) appears in the kinetic theory of membranes with enzymotic activity ([12]).
In this case $f_0(s) = \operatorname{sgn} s, \forall s \in \mathbb{R} \setminus \{0\}$. Since in applications one is essentially interested in non-negative solutions (u_ε^λ is interpreted as the dynamical concentration of enzyme in this case), one can also define $f(s)$ to be $f(s) = s_+(1+s)^{-1}$ with $s_+ = \max(s,0)$. Then one has $f_0(s) = \mathcal{H}(s)$.

Example 0.2. Let $1, a_+, a_- \in \mathbb{R}_+$ be fixed and let f_1 denote the piecewise linear function

$$f_1(s) = \ (s) \min\{1s,a_+\} + \ (-s) \max\{1s,-a_-\}$$

An asymptotic (for $\varepsilon \downarrow 0$) solution of $\mathfrak{T}_{\varepsilon,\infty}^\lambda$ with $f = f_1$ is constructed in section 1.2 below and used in order to investigate the asymptotic behaviour (for $\varepsilon \downarrow 0$) of the solution of $\mathfrak{T}_{\varepsilon,\infty}^\lambda$ with general f.

For a given function $u \in C^0(\bar{Q})$ (or $u \in C^0(\bar{U})$), denote by $E_+(u)$, $E_-(u)$ the sets where $u > 0$, $u = 0$, $u < 0$, respectively, whereas $\chi_+(u)$, $\chi_0(u)$, $\chi_-(u)$ stand for the characteristic functions of these sets.

We associate with $\mathcal{P}_\varepsilon^\lambda$ the following "reduced" problem $\mathcal{P}^\lambda$:

$$\begin{cases} \frac{\partial u^\lambda}{\partial t} + A(x,t,\frac{\partial}{\partial x})u^\lambda + \lambda f_0(u^\lambda) = g(x,t)\ [1-\chi_0(u^\lambda)], & (x,t) \in Q \\ \lambda f_{-\infty} \le g(x,t) \le \lambda f_{+\infty}, & (x,t) \in \text{int } E_0(u^\lambda) \\ u^\lambda(x,0) = \psi(x), & x \in \bar{U} \\ \pi_0 u^\lambda(x',t) = \phi(x',t), & (x',t) \in \Gamma. \end{cases} \tag{0.9}$$

The solution u^λ of (0.9) is supposed to be continuous in $\bar{Q}$, the differential equation and the condition $\lambda f_{-\infty} \le f(x,t) \le \lambda f_\infty$ in (0.9) are interpreted in the sense of Schwartz's distributions.

The corresponding stationary "reduced" problem $\mathcal{P}_\infty^\lambda$ is stated as follows :

$$\begin{cases} A_\infty(x,\frac{\partial}{\partial x})u_\infty^\lambda + \lambda f_0(u_\infty^\lambda) = g_\infty(x)\ [1-\chi_0(u_\infty^\lambda)]\ , & x \in U \\ \lambda f_{-\infty} \le g_\infty(x) \le \lambda f_{+\infty}, & x \in \text{int } (u_\infty^\lambda) \\ \pi_0 u_\infty^\lambda(x') = \phi_\infty(x'), & x' \in \partial U \end{cases} \tag{0.10}$$

where again $u_\infty^\lambda \in C^0(\bar{U})$, the differential equation and the condition $\lambda f_{-\infty} \le g_\infty(x) \le \lambda f_\infty$ in (0.10) are interpreted in the distributional sense. The reduced problems (0.9), (0.10) can be reformulated in terms of maximal monotone operators (see [3]).

I. STATIONARY PROBLEM.

1. General properties of the operators considered.

Both $\mathcal{P}_{\varepsilon,\infty}^\lambda$ and $\mathcal{P}_\infty^\lambda$ can be equivalently reformulated as variational minimization problems (see, for instance, [5]), where the corresponding functionals

$$D_{\varepsilon}^{\lambda}(u) = \int_U [\frac{1}{2} \sum_{1\le k,j\le n} a_{kj}^{\infty}(x) \frac{\partial u}{\partial x_k} \frac{\partial u}{\partial x_j} + \lambda\varepsilon F(\frac{u}{\varepsilon}) - g_{\infty} u]dx$$

$$D^{\lambda}(u) = \int_U [\frac{1}{2} \sum_{1\le k,j\le n} a_{kj}^{\infty}(x) \frac{\partial u}{\partial x_k} \frac{\partial u}{\partial x_j} + \lambda F_0(u) - g_{\infty} u]dx$$

are lower semi-continuous, coercive and strictly convex on the hyperplane

$$\Pi_{\Phi_{\infty}} = \{u \in H_1(U) \mid u - \Phi_{\infty} \in \overset{\circ}{H}_1(U)\}$$

Here, as usual, $H_1(U)$ is the Sobolev space of order 1 and $\overset{\circ}{H}_1(U)$ is the subspace of those functions in $H_1(U)$, for which traces on ∂U vanish ; further Φ_{∞} is the solution of the following boundary value problem :

$$\begin{cases} A_{\infty}(x,\frac{\partial}{\partial x})\Phi_{\infty} = g_{\infty}(x), & x \in U \\ \pi_0 \Phi_{\infty}(x') = \phi_{\infty}(x'), & x' \in \partial U. \end{cases} \tag{1.1.1}$$

Using the equivalent variational reformulation of $\mathfrak{J}_{\varepsilon,\infty}^{\lambda}$, $\mathfrak{J}_{\infty}^{\lambda}$ and the classical a priori estimates for linear second order elliptic operators, one gets the following result :

<u>Theorem 1.1.1</u>. *There exist well-defined solutions*

$$u_{\varepsilon,\infty}^{\lambda} \in C^{1,\alpha}(\bar{U}), \quad u_{\infty}^{\lambda} \in C^{1,\alpha}(\bar{U}), \ \forall \alpha \in [0,1)$$

of the problem $\mathfrak{J}_{\varepsilon,\infty}^{\lambda}$ *and* $\mathfrak{J}_{\infty}^{\lambda}$, *respectively.*

We use the same notation $\mathfrak{J}_{\varepsilon,\infty}^{\lambda}$ for the operator

$$\mathfrak{J}_{\varepsilon,\infty}^{\lambda} : H_2(U) \to L_2(U) \times H_{3/2}(\ U) \tag{1.1.2}$$

associated with the boundary value problem (0.8), where, as usual, $H_s(U)$ and $H_r(\partial U)$ stand for the Sobolev spaces (of orders s and r, respectively) of functions in U and on ∂U. We are going to state several results concerning the continuity properties of the nonlinear operator $\mathcal{T}^{\lambda}_{\varepsilon,\infty}$ and its inverse.

<u>Proposition 1.1.2</u>. *For any given* $\varepsilon > 0$, *the mapping (1.1.2) is a Lipschitz-continuous homeomorphism.*

<u>Proposition 1.1.3</u>. *For any given* $\phi_\infty \in H_{1/2}(\partial U)$, *the operator*

$$(\mathcal{T}^{\lambda}_{\varepsilon,\infty})^{-1} : H_{-1}(U) \to \Pi_{\Phi_\infty}$$

is Lipschitz-continuous uniformly with respect to ε.

<u>Proposition 1.1.4</u>. *With* $u^{\lambda}_{\varepsilon,\infty}$ *the solution of* $\mathcal{T}^{\lambda}_{\varepsilon,\infty}$, *the mapping*

$$\bar{R}_+ \ni \lambda \to u^{\lambda}_{\varepsilon,\infty} \in H_1(U)$$

is Lipschitz-continuous uniformly with respect to $\varepsilon \in (0,\varepsilon_0]$, *thus also for* $\varepsilon = 0$.

The proofs of Propositions 1.1.2 - 1.2.4 essentially use the fact that f(s) is monotonically increasing.

<u>Proposition 1.1.5</u>. *For any given* $g \in H_{-1}(U)$, *the operator*

$$H_{1/2}(\partial U) \ni \phi_\infty \xrightarrow{(\mathcal{T}^{\lambda}_{\varepsilon,\infty})^{-1}} u^{\lambda}_{\varepsilon,\infty} \in H_1(U) \tag{1.1.3}$$

is Hölder-continuous with exponent $\alpha = \frac{1}{2}$ *uniformly with respect to* ε. *Moreover, the following a priori estimate holds :*

$$\begin{aligned} &\|(\mathcal{T}^{\lambda}_{\varepsilon,\infty})^{-1}(\phi_1,g) - (\mathcal{T}^{\lambda}_{\varepsilon,\infty})^{-1}(\phi_2,g)\|_{H_1(U)} \le \\ &\le C(\|\Phi_1-\Phi_2\|^{1/2}_{L_2(U)} + \|\Phi_1-\Phi_2\|_{H_1(U)}), \end{aligned} \tag{1.1.4}$$

where Φ_j, $j=1,2$, *is the solution of the linear problem*

$$\begin{cases} A_\infty \Phi_j = 0, & x \in U, \\ \pi_0 \Phi_j = \phi_j, & x' \in \partial U \end{cases} \tag{1.1.5}$$

and the constant c *does not depend on* ϕ , g *and* ε.

Proof. Let $u_j = (\mathcal{T}^{\lambda}_{\varepsilon,\infty})^{-1}(\phi_j, g)$ and denote $v_j = u_j - \Phi_j$, where Φ_j is defined by (1.1.5). Then $v = v_1 - v_2$ is the solution of the problem :

$$\begin{cases} A_\infty v + \lambda[f(\frac{v_1+\Phi_1}{\varepsilon}) - f(\frac{v_2+\Phi_2}{\varepsilon})] = 0, & x \in U \\ \pi_0 v(x') = 0 & x' \in \partial U \end{cases} \tag{1.1.6}$$

The inner product in $L^2(U)$ after the integration by part yields :

$$\int_U \sum a^\infty_{kj}(x) v_{x_j} v_{x_k} + \lambda \int_U [f(\frac{v_1+\Phi_1}{\varepsilon}) - f(\frac{v_2+\Phi_2}{\varepsilon})](\Phi_1 - \Phi_2 + v)dx =$$

$$= \lambda \int_U f(\frac{v_1+\Phi_1}{\varepsilon}) - f(\frac{v_2+\Phi_2}{\varepsilon}) \; (\Phi_1 - \Phi_2)dx$$

The monotonicity of f(s) leads to the inequality :

$$\int_U \sum a^\infty_{kj}(x) v_{x_j} v_{x_k} dx \le \lambda \int_U [f(\frac{u_1}{\varepsilon}) - f(\frac{u_2}{\varepsilon})](\Phi_1 - \Phi_2)dx \le$$

$$\le 2\lambda \max\{f_{+\infty}, |f_{-\infty}|\} \; (\text{meas } U)^{1/2} \|\Phi_1 - \Phi_2\|_{L^2(U)} \tag{1.1.7}$$

As a consequence of (1.1.7), one gets the estimate :

$$\|\nabla v\|^2_{L_2(U)} \le 2\lambda\mu^{-1} \max\{f_{+\infty}, |f_{-\infty}|\}(\text{meas } U)^{1/2} \|\Phi_1 - \Phi_2\|_{L^2(U)}$$

where μ is the ellipticity constant for $A_\infty(x, \frac{\partial}{\partial x})$. Hence,

$$||v||_{H_1(U)} \leq C_1\|\Phi_1-\Phi_1\|^{1/2}_{L^2(U)}$$

where C_1 depends only on $\lambda,\mu,f_{\pm\infty}$, meas U and the constant ν in the Poincaré's inequality : $\|\nabla v\|_{L^2(U)} \leq \nu||v||_{L^2(U)}$, $\forall v \in \overset{\circ}{H}_1(U)$. As a consequence of (1.1.8), one gets (1.1.4) with $C = C_1 + 1$.

1.2. Convergence for $\varepsilon \to 0$.

In this section, an estimate for the H_1-norm of $u^\lambda_{\varepsilon,\infty} - u^\lambda_\infty$ is given which is a slight generalization of the result established previously in [16,2]. Moreover, if the free boundary $\partial E_0(u^\lambda_\infty)$ is a sufficiently smooth manifold, it is proved that the H_1-norm of $u^\lambda_{\varepsilon,\infty} - u^\lambda_\infty$ is of order $O(\varepsilon^{3/4})$ as $\varepsilon \to 0$. It will be shown in the next section that there exists a critical value $\lambda_c(\phi_\infty,g_\infty)$ of λ such that for $\lambda < \lambda_c$, the norm of $u^\lambda_{\varepsilon,\infty} - u^\lambda_\infty$ is of order $O(\varepsilon)$ as $\varepsilon \downarrow 0$.

Theorem 1.2.1. *Under the regularity assumption*

$$\phi_\infty \in C^2(\partial U), \quad g_\infty \in C^0(\bar{U}), \tag{1.2.1}$$

the following estimate holds :

$$\|u^\lambda_{\varepsilon,\infty}-u^\lambda_\infty\|_{H_1(U)} \leq C\varepsilon^{1/2}, \tag{1.2.2}$$

where the constant C *depends only on* λ,f,A_∞ *and* U.

Theorem 1.2.1. is proved by using the monotonicity of f and the fact that the functions $u^\lambda_{\varepsilon,\infty}$, u^λ_∞ can be characterized as solutions of corresponding elliptic variational inequalities.
Using a compactness argument and the uniqueness of the solution u^λ_∞ to the reduced problem, one can also prove the following

Proposition 1.2.2. *If* $g_\infty \in C^0(\bar{U})$, $\phi_\infty \in C^2(\partial U)$, *then one has :*

$$\lim_{\varepsilon\to 0} \|u^\lambda_{\varepsilon,\infty} - u^\lambda_\infty\|_{C^{1,\alpha}(\bar{U})} = 0, \qquad \alpha \in [0,1). \tag{1.2.3}$$

Now we are going to establish the main result in this section.

Theorem 1.2.3. *Assume that* $\phi_\infty > 0$ *on* ∂U, $g_\infty \equiv 0$ *in* U, *that the free boundary* $\partial E_0(u^\lambda_\infty)$ *is a* C^4*-manifold of dimension* $n-1$, *and that the function* f *satisfies the condition (0.2). Then the following estimate holds :*

$$\|u^\lambda_{\varepsilon,\infty} - u^\lambda_\infty\|_{H_1(U)} \leq C\varepsilon^{3/4}, \tag{1.2.4}$$

where the constant C *does not depend on* ε.

One constructs asymptotic solutions of the problem $\mathcal{T}^\lambda_{\varepsilon,\infty}$ with a specially chosen piecewise linear function $f = f_1$ in order to prove this theorem. More precisely, introducing the function

$$f_1(s) = \mathcal{H}(s)\min\{ls, f_\infty\} + \mathcal{H}(-s)\max\{ls, f_{-\infty}\} \tag{1.2.5}$$

where $\mathcal{H}(s)$ is Heaviside's function and the parameter $l > 0$ will be chosen later, one applies an appropriate modification of Vishik-Lyusternik's method ([19,20] ; see also [16,18]) for constructing asymptotic solutions of $\mathcal{T}^\lambda_{\varepsilon,\infty}$ with $f(s) = f_1(s)$ and for establishing (1.2.4).

For x in a sufficiently small neighbourhood of the free boundary $\partial E_0(U^\lambda_\infty)$, define $x' \in \partial E_0(u^\lambda_\infty)$, $\rho \in \mathbb{R}$ by the relations :

$$|x-x'| = \min_{y' \in \partial E_0(u^\lambda)} |x-y'| = \operatorname{dist}(x, \partial E_0(u^\lambda)) = |\rho|$$

$$\rho > 0 \quad \text{on} \quad E_+(u^\lambda)$$

$$\rho \leq 0 \quad \text{on} \quad E_0(u^\lambda).$$

If $x \in E_+(u^\lambda)$ and lies in the neighbourhood above (where the coordinates (x',ρ) are well defined), then the operator A_∞ can be rewritten as follows :

$$A_\infty(x,\frac{\partial}{\partial x}) = -\ (a(x)\ \frac{\partial^2}{\partial\rho^2} + (b(x)+c(x)\nabla')\ \frac{\partial^2}{\partial\rho^2} + B(x,\nabla'))$$

where ∇' denotes the gradient with respect to $x' \in \partial E_0(u^\lambda)$ and where $B(x,\nabla')$ is a differential operator of second order with sufficiently smooth coefficients. Besides, the functions $a(x) > 0$, $b(x)$, $c(x)$ are sufficiently smooth, since the manifold $\partial E_0(u^\lambda)$ is supposed to be sufficiently smooth. Let

$U_{\varepsilon,i} = \{x \in U \mid |\rho|_U < 2\varepsilon^{1/3}\}$ and $U_{\varepsilon,e} = \{x \in U \mid |\rho| > \varepsilon^{1/3}\}$ denote the interior and the exterior region, respectively. In $U_{\varepsilon,i}$, an asymptotic solution is sought in the form

$$v_\varepsilon(x',\rho) = \sum_{2\le j\le 3} \varepsilon^{j/2} v_j(x',\varepsilon^{-1/2}\rho), \tag{1.2.6}$$

where the functions v_j, $j=2,3$, are solutions of the following boundary value problems on $\mathbb{R}$ with $x' \in \partial E_0(u^\lambda)$ playing the role of a parameter :

$$\begin{cases} a(x')\ \dfrac{d^2}{d\zeta^2}\ v_2(x',\zeta) - \lambda f_1(v_2(x',\zeta)) = 0, & \zeta \in \mathbb{R} \\ v_2(x',\zeta) = O(1) & ,\quad \zeta \to -\infty \\ v_2(x',\zeta) = (2a(x'))^{-1}\lambda f_\infty \zeta^2 + O(1) & ,\quad \zeta \to +\infty \end{cases} \tag{1.2.7}$$

$$\begin{cases} a(x')\ \dfrac{d^2}{d\zeta^2}\ v_2(x',\zeta) - \lambda f_1'(v_2(x',\zeta))v_3(x',\zeta) = \gamma_3(x',\zeta), & \zeta \in \mathbb{R} \\ v_3(x',\zeta) = O(1) & ,\quad \zeta \to -\infty \\ v_3(x',\zeta) = \lambda f_\infty (3!)^{-1}\beta_3(x')\zeta^3 + O(1) & ,\quad \zeta \to +\infty \end{cases} \tag{1.2.8}$$

with

$$\gamma_3(x',\zeta) = -\ \lambda\zeta(a(x'))^{-1}a_\rho(x')f_1(x_2(x',\zeta)) - (b(x')+c(x')\nabla')v_{2\zeta}$$

$$\beta_3(x') = -\ (a(x'))^{-1}a_\rho(x') + (b(x')+c(x')\nabla')((a(x'))^{-1})$$

The solution of (1.2.7) is given by

$$v_2^{(1)}(x',\zeta) = \begin{cases} \frac{1}{1} e^{\left(\sqrt{\frac{\lambda f_\infty 1}{a(x')}}\zeta\right)-1}, & \zeta \le \sqrt{\frac{a(x')}{\lambda f_\infty 1}} \\ \frac{\lambda f_\infty}{2a(x')}\zeta^2 + \frac{1}{21}, & \zeta > \sqrt{\frac{a(x')}{\lambda f_\infty 1}} \end{cases}$$

and, $(d/d\zeta)v_2$ being a solution of the homogeneous differential equation in (1.2.18), the boundary value problem for v_3 can be solved using the variation of constants method.

Let $\chi \in C_0^\infty(\mathbb{R})$ be a function which is identically one on the interval $[-1,1]$ and the support of which is contained in $[-2,2]$. Let z_ε be defined by

$$z_\varepsilon(x) = \chi(\varepsilon^{-1/3}\rho)v_\varepsilon(x',\rho) + (1-\chi(\varepsilon^{-1/3}\rho))u_\infty^\lambda(x) \qquad (1.2.9)$$

Obviously, this function satisfies the boundary condition $\pi_0 z_\varepsilon = \phi_\infty$ on ∂U.

<u>Lemma 1.2.4</u>. *There exist constants* C, ε_0, *such that*

$$\|A_\infty z_\varepsilon + \lambda f_1(\frac{z_\varepsilon}{\varepsilon})\|_{L_2(U)} \le C\varepsilon^{5/6}$$

for $\varepsilon \in (0,\varepsilon_0]$.

<u>Proof</u>. We shall proceed by splitting the proof is several steps.

(i) For $x \in U_{\varepsilon,e}$, one has :

$$u^\lambda(x) \ge p\rho^2 \ge p\varepsilon^{2/3}$$

with a constant $p > 0$. Thus, $\varepsilon^{-1}z_\varepsilon(x) = \varepsilon^{-1}u\ (x) \ge p\varepsilon^{-1/3}$ and $f_1(\varepsilon^{-1}z_\varepsilon(x)) \equiv 1$ for $\varepsilon \in (0,\varepsilon_0]$, ε_0 sufficiently small. Thus, $A_\infty z_\varepsilon + \lambda f_1(\varepsilon^{-1}z_\varepsilon) \equiv 0$ for $\varepsilon \in (0,\varepsilon_0]$, $x \in U_{\varepsilon,e}$.

(ii) It will be shown that

$$\|f_1(v_2+\sqrt{\varepsilon}v_3)-(f_1(v_2)+\sqrt{\varepsilon}f_1'(v_2)v_3)\|_{L^2(U_{\varepsilon,i})} \le C\varepsilon \qquad (1.2.10)$$

where $v_j = v_j(x',\varepsilon^{-1/2}\rho)$. Without restriction of generality, one can assume that $\lambda = f_\infty = 1$. Since f_1 is piecewise linear, the function on the left hand side of (1.2.10) is zero if the interval $(v_2+\sqrt{\varepsilon}v_3, v_2)$ does not contain l^{-1}. Let now

$$S_\varepsilon = \{(x',\rho) \mid v_2 + \sqrt{\varepsilon}v_3 < l^{-1} < v_2\}.$$

If $(x',\rho) \in S_\varepsilon$, then $\varepsilon^{-1/2}\rho > \zeta_0(x') \overset{\text{def}}{=} \sqrt{l^{-1}a(x')}$ and

$$\frac{1}{2a(x')}\frac{\rho^2}{\varepsilon} + \frac{1}{2l} + \sqrt{\varepsilon}v_3 < l^{-1} < \frac{1}{2a(x')}\frac{\rho^2}{\varepsilon} + \frac{1}{2l}.$$

Since $|v_3(\zeta)| = O(\zeta^3)$, $\zeta \to \infty$, one obtains

$$\rho^2 - C\rho^3 < \varepsilon.\zeta_0(x')^2 < \rho^2$$

with a constant C. Thus, $|\rho-\sqrt{\varepsilon}\zeta_0(x')| \le C\varepsilon$. Since $l v_2(x',\zeta_0(x')) = 1$, the following inequality holds for $x \in S_\varepsilon$:

$$|f_1(v_2+\sqrt{\varepsilon}v_3)-(f_1(v_2)+\sqrt{\varepsilon}f_1'(v_2)v_3)| = |l(v_2+\sqrt{\varepsilon}v_3)-1| \le C\varepsilon^{1/2}.$$

The last inequality yields (1.2.10), since the measure of S_ε is of order $o(\varepsilon)$.

(iii) It will be shown that

$$\|A_\infty z_\varepsilon + \lambda(f_1(v_2)+f_1'(v_2)\sqrt{\varepsilon}v_3)\|_{C^0(\bar{U}_{\varepsilon,i})} \le C\varepsilon^{2/3} \qquad (1.2.11)$$

Consider first the region $|\rho| < \varepsilon^{1/3}$, where $\chi \equiv 1$. One has

$$B(x,\nabla')z_\varepsilon = B(x,\nabla')(\varepsilon v_2+\varepsilon^{3/2}v_3) = O(\varepsilon^{2/3}).$$

Thus, with $\zeta = \varepsilon^{-1/2}\rho$, one has

$$Az_\varepsilon + \lambda(f_1(v_2) + \sqrt{\varepsilon}f_1'(v_2)v_3) = r(x,\varepsilon) + O(\varepsilon^{2/3}),$$

where

$$\begin{aligned} r(x,\varepsilon) &= -a(x')v_{2\zeta\zeta}(x',\zeta) + \lambda f_1(v_2) + \\ &\quad + \sqrt{\varepsilon}(-a(x')v_{3\zeta\zeta}(x',\zeta) + \lambda f_1'(v_2)v_3 - \zeta a_\rho(x')v_{2\zeta\zeta}(x',\zeta) - \\ &\quad - (b(x') + c(x')\nabla')v_{2\zeta}(x',\tfrac{\rho}{\sqrt{\varepsilon}})) \\ &= O(\varepsilon^{2/3}) \end{aligned} \tag{1.2.12}$$

according to the construction of v_j.

In the region $\varepsilon^{1/3} < |\rho| < 2\varepsilon^{1/3}$, one obtains

$$Az_\varepsilon + \lambda(f_1(v_2)+\sqrt{\varepsilon}f_1'(v_2)v_3) = \sum_{0\le i\le 2} r_i(x,\varepsilon)$$

where r_i, $0 \le i \le 2$, are given by

$$\begin{aligned} r_0(x,\varepsilon) &= r(x,\varepsilon) - (a(x')+\rho a_\rho(x'))(1-\chi)(u_{\rho\rho} - (v_{2\zeta\zeta}+\sqrt{\varepsilon}v_{3\zeta\zeta})) \\ &\quad - (b+c\nabla')(1-\chi)(u_\rho - \sqrt{\varepsilon}v_{2\zeta}) + \chi B(\varepsilon v_2+\varepsilon^{3/2}v_3) + (1-\chi)Bu^\lambda \end{aligned}$$

$$\begin{aligned} r_1(x,\varepsilon) &= -\, 2a(x')\varepsilon^{-1/3}\chi'(\varepsilon^{-1/3}\rho)\frac{\partial}{\partial\rho}(\varepsilon v_2+\varepsilon^{3/2}v_3 - u^\lambda) \\ &\quad - (b+c\nabla')\varepsilon^{-1/3}\chi'(\varepsilon^{-1/3}\rho)(\varepsilon v_2+\varepsilon^{3/2}v_3 - u^\lambda) \end{aligned}$$

$$r_2(x,\varepsilon) = -a(x)\varepsilon^{-2/3}\chi''(\varepsilon^{-1/3}\rho)(\varepsilon v_2+\varepsilon^{3/2}v_3 - u^\lambda).$$

Using the boundary conditions for v_j and the asymptotic expansion

$$u^\lambda(x',\rho) = (2a(x')^{-1}\lambda f_\infty \rho^2 + \lambda f_\infty (3!)^{-1}\beta_3(x')\rho^3 + O(\rho^4), \quad \rho \downarrow 0,$$

one checks easily that

$$\sup_{\varepsilon^{1/3}<\rho<2\varepsilon^{1/3}} |r_i(x,\varepsilon)| \le C\varepsilon^{2/3}, \quad i=0,1,2,$$

where the constant C does not depend upon ε. Thus (1.2.11) is proved.

(iv) Since $u^\lambda(x) \gg p\rho^2$ for $x \in E_+(u^\lambda)$ with a constant $p > 0$, one can choose ε_0 so small that for $\forall \varepsilon \in (0,\varepsilon_0]$, $\forall x \in U_{\varepsilon,e}$, $f(\varepsilon^{-1}z_\varepsilon(x)) = f_\infty$ and $f(v_2(x',\varepsilon^{-1/2}\rho) + \varepsilon^{1/2}v_3(x',\varepsilon^{-1/2}\rho)) = f_\infty$ for $x \in U_{\varepsilon,i} \cap U_{\varepsilon,e}$. Thus

$$\|A_\infty z_\varepsilon(x) + \lambda f_1(\varepsilon^{-1}z_\varepsilon(x))\|_{L_2(U)} =$$

$$= \|A_\infty z_\varepsilon(x) + \lambda f_1(\varepsilon^{-1}z_\varepsilon(x))\|_{L_2(U_{\varepsilon,i})}$$

$$= \|A_\infty z_\varepsilon(x) + \lambda f_1(v_2+\varepsilon^{1/2}v_3)\|_{L_2(U_{\varepsilon,i})}$$

$$\|A_\infty z_\varepsilon + \lambda(f_1(v_2)+f_1'(v_2)\sqrt{\varepsilon}v_3)\|_{L_2(U_{\varepsilon,i})}$$

$$+ \lambda\|f_1(v_2) + f_1'(v_2)\varepsilon^{1/2}v_3 - f_1(v_2+\varepsilon^{1/2}v_3)\|_{L_2(U_{\varepsilon,i})}$$

$$\le C\varepsilon^{5/6},$$

as a consequence of (1.2.10), (1.2.11) and given that meas $(U_{\varepsilon,i}) = O(\varepsilon^{1/3})$, $\varepsilon \downarrow 0$.

Lemma 1.2.4 is proved. □

In order to prove Theorem 1.2.3 above, several auxiliary results will be needed.

<u>Lemma 1.2.5</u>. *There exists a well defined value of the parameter* $\in (0,\infty)$, *such that*

$$I(l) = \int_{-\infty}^{\infty} (f(v_2^{(l)}(\zeta)) - f_1(v_2^{(l)}(\zeta)))d\zeta = 0 \tag{1.2.13}$$

<u>Proof</u>. Let $v(\zeta) = v_2^{(l)}(\zeta)$. Then $v_2^{(l)}(\zeta) = l^{-1}v(l^{1/2}\zeta)$. Using the substitution $\eta = v(\sqrt{l}\zeta)$, one gets on

$$\sqrt{l}I(l) = \int_0^{\infty} (f(l^{-1}\eta)-f_1(\eta)) \frac{d\eta}{v'(v^{-1}(\eta))}$$

The right hand side is a strictly decreasing function of $l \in (0,\infty)$, so that $I(l)$ has at most one zero. For l sufficiently large, one has $f(l^{-1}\eta) - f_1(\eta) < 0 \quad \eta > 0$, so that $I(l) < 0$ for $l \gg 1$. Now $I(l) > 0$ for l sufficiently small. In fact, one has :

$$\begin{aligned} &|\int_1^{\infty} (f(l^{-1}\eta)-f_1(\eta))(v^{-1}(\eta)))^{-1}d\eta| \\ &= \int_1^{\infty} (f(l^{-1}\eta)-f_\infty)(v'(v^{-1}(\eta)))^{-1}d\eta| \\ &\leq \; Cl \int_1^{\infty} \eta^{-3/2}d\eta \end{aligned} \tag{1.2.14}$$

with C independent of l. Hence,

$$\sqrt{l}I(l) = \int_0^1 (f(l^{-1}\eta)-f_1(\eta))(v'(v^{-1}(\eta)))^{-1}d\eta + O(l), \text{ when } l \downarrow 0.$$

Therefore,

$$\lim_{l\downarrow 0} \sqrt{l}I(l) = \int_0^1 (f_\infty-f_1(\eta))(v'(v^{-1}(\eta)))^{-1}d\eta > 0,$$

and lemma 1.2.5 is proved.

We choose l to be the zero of $I(l)$. Let $h(s) = f(s) - f_1(s)$, and for $a > 0$, define $U_a = \{x \in U \mid \text{dist}(x,\partial E_0(u_\infty^\lambda)) > a\}$ $R_a = \{\rho \in \mathbb{R} \mid |\rho| > a$.

We choose $a > 0$ so small that for $|\rho| < a$, the mapping $x \to (x',\rho)$ is a diffeomorphism.

Lemma 1.2.6. *There exist constants* $C,\varepsilon_0 > 0$, *which do not depend upon* ε *and such that for* $\varepsilon \in (0,\varepsilon_0]$ *holds :*

(i) $\|h(\varepsilon^{-1}z_\varepsilon(x)\|_{L^2(U_a)} \le C\varepsilon$

(ii) $\|h(v_2(x',\varepsilon^{-1/2}\rho))\|_{L^2(\partial E_0 \times \mathbb{R}_a)} \le C\varepsilon$

(iii) $\|h(v_2(x',\varepsilon^{-1/2}\rho))\|_{H_{-1}(U\setminus U_a)} \le C\varepsilon^{3/4}$.

Proof.(i) For $x \in U_a \cap E_+(u_\infty^\lambda)$, one has : $z_\varepsilon(x) = u^\lambda(x) \ge p > 0$, where p does not depend on ε,x. The inequality $|h(s)| \le C(1+|s|)^{-1}$ and the fact that for $x \in U_a \cap E_0(u_\infty^\lambda)$, one has : $z_\varepsilon(x) \equiv 0$, so that $h(\varepsilon^{-1}z_\varepsilon(x)) = 0$ yield the first part of the Lemma.

(ii) Since for $\rho \in \mathbb{R}_a$ holds $v_2(x',\varepsilon^{-1/2}\rho) \ge \varepsilon^{-1}p$, where p does not depend on x',ε, the second inequality can be proved similarly to the first one.

(iii) Denote $t(x',s) = h(v_2(x',s))$. One has

$$\|t(x',\varepsilon^{-1/2}\rho)\|_{H_{-1}(\partial E_0\times(-a,0))} \le$$

$$\le \|t(x',\varepsilon^{-1/2}\rho)\|_{H_{-1}(\partial E_0 \times \mathbb{R})} + \|t(x',\varepsilon^{-1/2}\rho)\|_{L^2(\partial E_0 \times \mathbb{R}_a)} \qquad (1.2.15)$$

Let $T(x',\zeta) = \int_{-\infty}^{\zeta} t(x',s)ds$. The inequality $|t(x',s)| \le C(1+s^2)^{-1}$, which holds uniformly with respect to $x' \in \partial E_0$, and Lemma 1.2.5 yield $|T(x',\zeta)| \le C(1+|\zeta|)^{-1}$. Thus, $T \in L^2(\partial E_0 \times \mathbb{R})$ and

$$\|t(x',\varepsilon^{-1/2}\rho)\|_{H_{-1}(\partial E_o \times \mathbb{R})} = \sup_{\phi\in H_1(U)} \|\phi\|^{-1}_{H_1(U)} \left|\int t(x',\varepsilon^{-1/2}\rho)\phi(x',\rho)dx'd\rho\right|$$

$$= \sup_{\phi\in H_1} \|\phi\|^{-1}_{H_1(U)} \left|\int \varepsilon^{1/2} T(x',\varepsilon^{-1/2}\rho)\ \frac{\partial\phi}{\partial\rho}(x',\rho)dx'd\rho\right|$$

$$\leq \varepsilon^{1/2}\|T(x',\varepsilon^{-1/2}\rho)\|_{L^2(\partial E_o \times \mathbb{R})} \sup_{\phi\in H_1(U)} \|\phi\|^{-1}_{H_1(U)} \left\|\frac{\partial\phi}{\partial\rho}\right\|_{L^2(U)}$$

$$\leq C\varepsilon^{3/4}.$$

As a consequence of (ii), the second term on the right hand side of (1.2.15) is of order $O(\varepsilon)$ and that ends the proof of Lemma 1.2.6. □

Lemma 1.2.7. (i) *There exist constants* C, ε_o, *such that*

$$C^{-1}|v_2(\varepsilon^{-1/2}\rho)| \leq \varepsilon^{-1}|z_\varepsilon(x)| \leq C|v_2(\varepsilon^{-1/2}\rho)| \quad \forall\rho \in (0,a),\ \forall\varepsilon \in (0,\varepsilon_o]. \tag{1.2.16}$$

(ii) *There exists a constant* $C > 0$ *such that*

$$\|h(\varepsilon^{-1}z_\varepsilon(x))-h(v_2(\varepsilon^{-1/2}\rho)\|_{L^2(U\setminus U_a)} \leq C\varepsilon^{3/4} \quad \forall\varepsilon \in (0,\varepsilon_o]. \tag{1.2.17}$$

Proof. (i) For $0 < \rho < \varepsilon^{1/3}$, one has

$$\varepsilon^{-1}z_\varepsilon(x) = v_2(\varepsilon^{-1/2}\rho)(1+(v_2(\varepsilon^{-1/2}\rho))^{-1}\varepsilon^{1/2}v_3(\varepsilon^{-1/2}\rho)).$$

The inequality

$$|v_2(\varepsilon^{-1}\rho)^{-1}\varepsilon^{1/2}v_3(\varepsilon^{-1/2}\rho)| \leq C\rho \leq C\varepsilon^{1/3}$$

implies that (1.2.16) holds for $0 < \rho < \varepsilon^{1/3}$. Now let $\varepsilon^{1/3} < \rho < 2\varepsilon^{1/3}$. Then

$$\varepsilon^{-1}z_\varepsilon(x) = v_2(\varepsilon^{-1/2}\rho)(1+r_1(\varepsilon,x))$$

where the function

$$r_1(\varepsilon,x) = v_2(\varepsilon^{-1/2}\rho)^{-1}[\chi(\varepsilon^{-1/3}\rho)\varepsilon^{1/2}v_3 + (1-\chi(\varepsilon^{-1/3}\rho)(\varepsilon^{-1}u^\lambda(x)-v_2(\varepsilon^{-1/2}\rho))]$$

can be estimated as follows :

$$|r_1(\varepsilon,x)| \leq C\rho \leq 2C\varepsilon^{1/3} \qquad \forall\varepsilon \in (0,\varepsilon_0],$$

with some constant $C > 0$.

Finally, let $2\varepsilon^{1/3} < \rho < a$. Then

$$\varepsilon^{-1}z_\varepsilon(x) = \varepsilon^{-1}u^\lambda(x) = \varepsilon^{-1}((2a(x'))^{-1}\rho^2 + o(\rho^3))$$
$$= v_2(\varepsilon^{-1/2}\rho)(1+o((v_2(\varepsilon^{-\frac{1}{2}}\rho))^{-1}))$$

and (1.2.16) is proved.

(ii) The left hand side of (1.2.17) can be estimated as follows :

$$\|h(\varepsilon^{-1}z_\varepsilon(x))-h(v_2(\varepsilon^{-1/2}\rho)\|^2_{L^2(U\ U_a)} \leq$$

$$\leq \int_{|\rho|< a} (\sup_{\theta\in(v_2,\varepsilon^{-1}z_\varepsilon)} |h'(\theta)|)^2|\varepsilon^{-1}z_\varepsilon(x)-v_2(\varepsilon^{-1/2}\rho)|^2dx$$

$$\leq C \int_{0<\rho<a} (\sup_{\theta\in(v_2,\varepsilon^{-1}z_\varepsilon)} (1+\theta^2)^{-1})^2|\varepsilon^{-1}z_\varepsilon(x)-v_2(\varepsilon^{-1/2}\rho)|^2dx +$$

$$+ C\int_{-a<\rho<0}|\varepsilon^{-1}z_\varepsilon(x)-v_2(\varepsilon^{-1/2}\rho)|^2dx.$$

Using (1.2.16) and the asymptotic behaviour of $v_2(\zeta)$ for $\zeta \to \infty$, one obtains :

$$\|h(\varepsilon^{-1}z_\varepsilon(x))-h(v_2(\varepsilon^{-1/2}\rho))\|^2_{L^2(U\setminus U_a)} \le$$

$$\le C\int_{0<\rho<a}(1+(\varepsilon^{-1/2}\rho)^4)^{-2}(1+\varepsilon^{-1}\rho^3)^2 d\rho + C\int_{-a<\rho<0}\varepsilon|v_3(\varepsilon^{-1/2}\rho)|^2 d\rho$$

$$\le C\varepsilon^{3/2} \quad \forall\varepsilon \in (0,\varepsilon_0],$$

where C does not depend upon ε, and that ends the proof of Lemma 1.2.7.

Proof of Theorem 1.2.3. Let $l \in (0,\infty)$ be the zero of the function $I(l)$ defined by (1.2.13), and let $z_\varepsilon(x)$ be given by (1.2.9). One has :

$$A_\infty z_\varepsilon + \lambda f_l(\varepsilon^{-1}z_\varepsilon(x)) = r_\varepsilon(x),$$

where, according to Lemma 1.2.4,

$$\|r_\varepsilon\|_{L_2(U)} \le C\varepsilon^{5/6}, \qquad \varepsilon \in (0,\varepsilon_0], \tag{1.2.18}$$

with some constant $C > 0$ which does not depend on ε.

Writing the differential equations for u_ε^λ and for z_ε and taking the difference, on gets for $u_\varepsilon^\lambda - z_\varepsilon$ the following differential equation.

$$A_\infty(u_\varepsilon^\lambda - z_\varepsilon) + \lambda(f(\varepsilon^{-1}u_\varepsilon^\lambda(x))-f(\varepsilon^{-1}z_\varepsilon(x))) =$$

$$= - r_\varepsilon(x) - \lambda(f(\varepsilon^{-1}z_\varepsilon(x))-f_l(\varepsilon^{-1}z_\varepsilon(x)))$$

Taking the inner product with $u_\varepsilon^\lambda - z_\varepsilon$ in $L^2(U)$ in the last equation and using the monotonicity of f, one gets the following estimates :

$$\int_U (u_\varepsilon^\lambda - z_\varepsilon)A_\infty(u_\varepsilon^\lambda - z_\varepsilon)dx \le (\|r_\varepsilon\|_{H_{-1}(U)} + \lambda\|f(\varepsilon^{-1}z_\varepsilon(x))-f_l(\varepsilon^{-1}z_\varepsilon(x))\|_{H_{-1}(U)})$$

$$\cdot\ \|u_\varepsilon^\lambda - z_\varepsilon\|_{H_1(U)}$$

The integration by part and Poincaré's Lemma yield :

$$\|u_\varepsilon^\lambda - z_\varepsilon\|_{H_1(U)} \le C(\|r_\varepsilon\|_{H_{-1}(U)} + \|f(\varepsilon^{-1}z_\varepsilon(x)) - f_1(\varepsilon^{-1}z_\varepsilon(x)\|_{H_{-1}(U)})$$

The last inequality is also a consequence of Proposition 1.1.3.

According to (1.2.18), the first term on the right hand side is bounded by $C\varepsilon^{5/6}$.

According to the Lemmas 1.2.6, 1.2.7, the second term can be estimated as follows :

$$\begin{aligned}
&\|f(\varepsilon^{-1}z_\varepsilon(x)) - f_1(\varepsilon^{-1}z_\varepsilon(x)\|_{H_{-1}(U)} \\
&= \|h(\varepsilon^{-1}z_\varepsilon(x))\|_{H_{-1}(U)} \\
&\le \|h(\varepsilon^{-1}z_\varepsilon(x))\|_{H_{-1}(U_a)} + \|h(\varepsilon^{-1}z_\varepsilon(x))\|_{H_{-1}(U\setminus U_a)} \\
&\le \|h(\varepsilon^{-1}z_\varepsilon(x))\|_{L^2(U_a)} + \|h(v_2(x',\varepsilon^{-1/2}))\|_{H_{-1}(U\setminus U_a)} + \\
&\qquad + \|h(\varepsilon^{-1}z_\varepsilon(x)) - h(v_2(x',\varepsilon^{-1/2}))\|_{H_{-1}(U\setminus U_a)} \\
&\le C\varepsilon^{3/4}.
\end{aligned}$$

This ends the proof of Theorem 1.2.3. □

<u>Remark 1.2.8</u>. Consider the problem $\mathfrak{I}_{\varepsilon,\infty}^\lambda$ in $U = (-1,1) \subset \mathbb{R}$ with $f = f_1$ defined in (1.2.5) and $A_\infty = -(\frac{d}{dx})^2$, $g_\infty \equiv 0$, $\phi_\infty(x') = 1$. The function $u_\infty^\lambda(x) = (\lambda/2)(|x|-\xi)_+^2$, $\xi = 1 - (2\lambda^{-1})^{1/2}$, is the solution of the reduced problem $\mathfrak{I}_\infty^\lambda$. Let v_2 be the solution of (1.2.7) and let

$$z_\varepsilon(x) = u_\infty^\lambda(x)(1-\chi(|x|-\xi)) + \varepsilon v_2(\sqrt{\lambda/\varepsilon}(|x|-\xi))\chi(|x|-\xi). \qquad (1.2.19)$$

Similarly to the proof of Lemma 1.2.4, one checks that

$$\|-z_\varepsilon'' + \lambda f_1(\varepsilon^{-1}z_\varepsilon)\|_{C^0(\bar{U})} \le C\varepsilon \qquad (1.2.20)$$

where the constant C does not depend on ε. Further, $\pi_o z_\varepsilon = 1$. Partial integration yields :

$$\|u^\lambda_{\varepsilon,\infty} - z_\varepsilon\|_{H_1(U)} \le C\varepsilon, \tag{1.2.21}$$

where the constant C does not depend on ε.

Since $f_1'(+0) > 0$, the function $v_2(\zeta)$ and its derivative decrease exponentially for $\zeta \to -\infty$. Thus,

$$\gamma.\varepsilon^{3/4} \le \|z_\varepsilon\|_{H_1(-\xi,\xi)} \quad \forall\, \varepsilon \in (0,\varepsilon_o], \tag{1.2.22}$$

where the constant $\gamma > 0$ does not depend on ε. The inequalities (1.2.21), (1.2.22) yield :

$$\|u^\lambda_{\varepsilon,\infty} - u^\lambda_\infty\|_{H_1(U)} \ge \|u^\lambda_{\varepsilon,\infty} - u^\lambda_\infty\|_{H_1(-\xi,\xi)}$$

$$\ge \|u^\lambda_{\varepsilon,\infty}\|_{H_1(-\xi,\xi)} \ge \|z_\varepsilon\|_{H_1(-\xi,\xi)} - \|u^\lambda_{\varepsilon,\infty} - z_\varepsilon\|_{H_1(U)}$$

$$\ge \gamma\varepsilon^{3/4} - C\varepsilon$$

$$\ge \frac{\gamma}{2}\,\varepsilon^{3/4}, \quad \forall\, \varepsilon \in (0,\varepsilon_o],$$

where $\varepsilon_o > 0$ is sufficiently small. Thus, one has the following two-sided error estimate :

$$C^{-1}\varepsilon^{3/4} \le \|u^\lambda_{\varepsilon,\infty} - u^\lambda_\infty\|_{H_1(U)} \le C\varepsilon^{3/4} \tag{1.2.23}$$

with some constant $C > 0$ which does not depend on ε. It can be shown that the estimate (1.2.23) holds in the general case, as well, if the assumptions of Theorem 1.2.3 and the condition $f'(+0) > 0$ are satisfied.

Remark 1.2.9. If $g_\infty \not\equiv 0$, then the same argument with corresponding slight modifications in the construction of the asymptotic solutions of the problem $\mathcal{P}^\lambda_{\varepsilon,\infty}$ with $f = f_1$ leads to the same estimate under the assumptions of

Theorem 1.2.3.

1.3. The critical value λ_c of λ.

If $\phi_\infty(x') \neq 0$ $\forall x' \in \partial U$, then some critical value λ_c of the parameter λ plays a special role in the investigation of the boundary value problem $\mathfrak{I}_\infty^\lambda$. Namely, if $\phi_\infty > 0$, then for $\lambda < \lambda_c$ the problem $\mathfrak{I}_\infty^\lambda$ becomes linear, whereas for $\lambda > \lambda_c$ it is a nonlinear problem with piecewise constant discontinuous (across $\partial E_0(u_\infty^\lambda)$) nonlinearity. Denote by $C_+^0(\partial U)$ and $C_-^0(\partial U)$ the cones of continuous positive and negative functions on ∂U, respectively. Further, let $G(x,y)$ be Green's function for $A_\infty(x,\frac{\partial}{\partial x})$ in U with Dirichlet boundary conditions on ∂U and denote by $E(x,y')$ the Poisson kernel for the Dirichlet problem for the equation $A_\infty u = 0$.

Theorem 1.3.1. (i) *If* $\phi_\infty \in C_+^0(\partial U)$ *(respectively,* $\phi_\infty \in C_-^0(\partial U)$*), then there exists a well defined critical value* $\lambda_c = \lambda_c^+(\phi_\infty, g_\infty)$ *(respectively,* $\lambda_c = \lambda_c^-(\phi_\infty, g_\infty)$*), such that*

$$E_0(u_\infty^\lambda) = \emptyset \text{ if } \lambda < \lambda_c,$$

meas $(E_0(u_\infty^\lambda) \cup E_-(u_\infty^\lambda)) > 0$ *if* $\lambda > \lambda_c$ *(respectively, meas*$(E_0(u_\infty^\lambda) \cup E_+(u_\infty^\lambda)) > 0$ *if* $\lambda > \lambda_c$*)*

(ii) *If* $\lambda_c \geq \lambda$, *then* u_∞^λ *is the solution of the linear problem*

$$\begin{aligned} &A_\infty(x,\tfrac{\partial}{\partial x})v_\infty^\lambda(x) = g_\infty(x) - \lambda f_\infty, && x \in U \\ &\pi_0 v_\infty^\lambda(x) = \phi_\infty(x'), && x' \in \partial U \end{aligned} \tag{1.3.1}$$

If $g_\infty \equiv 0$, *then meas* $E_0(u_\infty^{\lambda_c}) = 0$.

(iii) *The functionals* λ_c^+, λ_c^- *can be represented as follows :*

$$\lambda_c^{\pm}(\phi_\infty, g_\infty) = \min_{x \in U} \Lambda^{\pm}_{\phi_\infty, g_\infty}(x), \tag{1.3.2}$$

where the function $\Lambda^{\pm}_{\phi_\infty, g_\infty}$ *is defined by :*

$$\Lambda^{\pm}_{\phi_\infty,g_\infty}(x) = (f_{\pm\infty}\int_U G(x,y)dy)^{-1}(\int_{\partial U} E(x,y')\phi_\infty(y')d\sigma_{y'} + \int_U G(x,y)g\ (y)dy)$$

The proof of Theorem 1.3.1 is similar to the proof of Theorem 2 in [9].

Proposition 1.3.2. *The functional* $(\phi_\infty,g_\infty) \to \lambda_c^+(\phi_\infty,g_\infty)$ *has the following properties* :

(i) $\lambda_c^+(\alpha\phi_\infty,\alpha g_\infty) = \alpha\lambda_c^+(\phi_\infty,g_\infty), \ \forall\, \alpha > 0$

(ii) $\lambda_c^+(\phi_\infty^{(1)},g_\infty^{(1)}) \le \lambda_c^+(\phi_\infty^{(2)},g_\infty^{(2)}), \quad 0 \le \phi_\infty^{(1)} \le \phi_\infty^{(2)},\ g_\infty^{(1)} \le g_\infty^{(2)}$

(iii) $\lambda_c^+(\gamma\phi_\infty^{(1)}+(1-\gamma)\phi_\infty^{(2)},\ \ g^{(1)}+(1-\gamma)g_\infty^{(2)}) \ge \gamma\lambda_c^+(\phi_\infty^{(1)},g_\infty^{(1)}) +$

$+ (1-\gamma)\lambda_c^+(\phi_\infty^{(2)},g_\infty^{(2)}),\ \forall\gamma \in [0,1].$

Proof. One proves (i) - (iii), using the formula (1.3.2) for $\lambda_c^+(\phi_\infty,g_\infty)$. □

Analogous properties has the functional $\lambda_c^-(\phi_\infty,g_\infty)$.

For $\lambda < \lambda_c$, the convergence result given in section 1.2 can be improved. One has the

Theorem 1.3.3. *Assume that* $\phi_\infty \in C_+^0(\partial U)$ *and that* $\lambda_c > \lambda$. *Then the following estimate holds :*

$$\|u^\lambda_{\varepsilon,\infty}-u^\lambda_\infty\|_{W_{2,p}(U)} \le C\varepsilon(\lambda_c-\lambda)^{-1}, \quad \forall\varepsilon > 0,\ \forall p, 1 < p < \infty \tag{1.3.4}$$

where the constant C *depends only upon* $p,L,U,A_\infty,\phi_\infty$ *and* g_∞.

Proof. Since $\mathcal{T}^\lambda_\infty$ for $\lambda < \lambda_c$ is linear, the function $w = u^\lambda_{\varepsilon,\infty}-u^\lambda_\infty$ is the solution of the problem :

$$\begin{cases} A_\infty w(x) = \lambda[f_{+\infty} - f(\frac{u^\lambda_{\varepsilon,\infty}}{\varepsilon})]\ , & x \in U \\ \pi_0 w(x') = 0, & x' \in \partial U. \end{cases} \tag{1.3.5}$$

One can write :

$$u_\infty^\lambda(x) = u^{\lambda_c}(x) + (\lambda_c-\lambda)v(x), \ \lambda \le \lambda_c$$

where v(x) is the solution of the problem

$$\begin{cases} A_\infty v(x) = 1, & x \in U \\ \pi_0 v(x') = 0, & x' \in \partial U \end{cases} \qquad (1.3.6)$$

Since $v(x) > 0, \forall x \in U$ and $u^{\lambda_c}(x') > 0, \forall x' \in \partial U$, one can find a constant $\gamma > 0$ such that the sets :

$$U_1 = \{x \in U \mid u^{\lambda_c}(x) \ge \gamma(\lambda_c-\lambda)\}, \quad U_2 = \{x \in U \mid v(x) \ge \gamma(\lambda_c-\lambda)\}$$

cover $\bar{U}$, so that one has :

$$u_\infty^\lambda(x) \ge \gamma(\lambda_c-\lambda), \quad \forall x \in \bar{U}$$

Of course, γ depends upon ϕ_∞ and g_∞.

Denote $v_\varepsilon(x) = \dfrac{\partial u_{\varepsilon,\infty}^\lambda}{\partial \varepsilon}$. Then v_ε is the solution of the problem :

$$\begin{cases} A_\infty v_\varepsilon + \frac{\lambda}{\varepsilon} f'\left(\frac{u_{\varepsilon,\infty}^\lambda}{\varepsilon}\right) v_\varepsilon = \frac{1}{\varepsilon^2} f'\left(\frac{u_{\varepsilon,\infty}^\lambda}{\varepsilon}\right) u_{\varepsilon,\infty}, & x \in U \\ \pi_0 v_\varepsilon(x') = 0, & x' \in U \end{cases}$$

Since $f'(s) \ge 0$, $u_{\varepsilon,\infty}^\lambda \ge 0$ for ε sufficiently small (because $u_\infty^\lambda > 0$, $\forall x \in \bar{U}, \forall \lambda < \lambda_c$), one gets the conclusion that $v_\varepsilon(x) \ge 0$, so that $u_{\varepsilon,\infty}^\lambda$ is monotonically increasing function of ε and, in particular, one has :

$$u_{\varepsilon,\infty}^\lambda(x) \ge u_\infty^\lambda(x) \ge \gamma(\lambda_c-\lambda), \quad \forall x \in \bar{U}, \quad \varepsilon > 0$$

Hence,

$$0 \le f_{+\infty} - f\left(\frac{u^{\lambda}_{\varepsilon,\infty}}{\varepsilon}\right) = \int_{\frac{u^{\lambda}_{\varepsilon,\infty}}{\varepsilon}}^{\infty} f'(s)ds \le L \frac{\varepsilon}{\varepsilon + u^{\lambda}_{\varepsilon,\infty}} \le \frac{L}{\gamma(\lambda_c - \lambda)} \varepsilon$$

and

$$\| f_{+\infty} - f\left(\frac{u^{\lambda}_{\varepsilon,\infty}}{\varepsilon}\right) \|_{L_p(U)} \le \frac{L}{\gamma} \varepsilon (\lambda_c - \lambda)^{-1} \quad (\text{meas } U)^{1/p}. \tag{1.3.7}$$

As a consequence of the a priori estimates for second order linear elliptic operators, one gets, using (1.3.5), (1.3.6), the estimate (1.3.4).

Corollary 1.3.4. *If* $\lambda_c - \lambda = \varepsilon^{\theta}$, $0 < \theta < 1$, *then* $u^{\lambda}_{\varepsilon,\infty}$ *converges to* $u^{\lambda_c}_{\infty}$ *in* $W_{2,p}(U)$ $\forall p < \infty$, *as* $\varepsilon \to 0$,

Corollary 1.3.5. *If* $\lambda_c - \lambda = \varepsilon^{\theta}$, $0 < \theta < 1$, *then* $u^{\lambda}_{\varepsilon,\infty}$ *converges to* $u^{\lambda_c}_{\infty}$ *in* $H_1(U)$ *and the rate of convergence is* $O(\varepsilon^{\theta})$. *In fact, let* $\mu = \lambda_c - \lambda^{1-\theta}$. *Then, using Proposition 1.1.4 and Theorem 1.3.3, one gets :*

$$\begin{aligned} \| u^{\lambda}_{\varepsilon,\infty} - u^{\lambda_c}_{\infty} \|_1 &\le \| u^{\lambda}_{\varepsilon,\infty} - u^{\mu}_{\varepsilon,\infty} \|_1 + \| u^{\mu}_{\varepsilon,\infty} - u^{\lambda_c}_{\infty} \|_1 \\ &\le C(|\lambda - \mu| + C\varepsilon(\lambda_c - \mu)^{-1}) \\ &\le C\,\varepsilon^{\theta}. \end{aligned}$$

1.4. Estimates for the critical value of λ.

Let $G(x,y)$ denote Green's function for the operator $A\,(x,\frac{\partial}{\partial x})$ with homogeneous Dirichlet boundary conditons and let $E(x,y')$ be the Poisson kernel for the Dirichlet problem for the equation $A_{\infty} u = 0$. Denote by $C^{o}_{+}(U)$ and $C^{o}_{+}(\partial U)$ the cone of positive continuous functions on U and ∂U, respectively. For functions $g \in C^{o}_{+}(U)$ and $\phi \in C^{o}_{+}(\partial U)$, let $\mathcal{M}_t(g)$ and $\mathcal{M}_t(\phi)$ be their mean values of order t :

$$\mathcal{M}_t(g) = \left(\frac{1}{\text{meas}(U)}\int_U g(x)^t dx\right)^{\frac{1}{t}}, \qquad \forall t \in \mathbb{R}$$

$$\mathcal{M}_t(\phi) = \left(\frac{1}{\text{meas}(\partial U)}\int_{\partial U} \phi(x')^t d\sigma_{x'}\right)^{\frac{1}{t}}, \quad \forall t \in \mathbb{R}$$

$$\mathcal{M}_t(\phi) = \left(\frac{1}{2}\sum_{x'\in\partial U} \phi(x')^t\right)^{\frac{1}{t}}, \qquad \forall t \in \mathbb{R}, \text{ if } n=1.$$

In this section, it is assumed that $g_\infty \in C^0_+(U)$, $\phi_\infty \in C^0_+(\ U)$. For the investigation of the functional $\lambda_c(\phi_\infty,g_\infty)$, the assumption of positivity of g_∞ is not a restriction of the generality. Indeed, (1.3.2) implies that :

$$\lambda^+_c(\phi_\infty,g_\infty) = \lambda^+_c(\phi_\infty,g_\infty+\rho)-(f_\infty)\rho^{-1}, \qquad \forall\rho \in \mathbb{R}.$$

Let

$$v(x) = \int_U G(x,y)dy \tag{1.4.1}$$

and let $\sum$ be the set of the points where the function v attains a global maximum. For functions $g \in C^0_+(U)$, $\phi \in C^0_+(\partial U)$ and for $x_0 \in \sum$, define the mean values $\mathcal{N}_{x_0}(g)$, $\mathcal{N}_{x_0}(\phi)$ as follows :

$$\mathcal{N}_{x_0}(g) = (v(x_0))^{-1}\int_U G(x_0,y)g(y)dy,$$

$$\mathcal{N}_{x_0}(\phi) = \int_{\partial U} E(x_0,y')\phi(y')d\sigma_{y'}.$$

One has the

Proposition 1.4.1. *Let* $U \subset \mathbb{R}^n$, $n \geq 2$. *For* $\forall(\phi_\infty,g_\infty) \in C^0_+(\partial U) \times C^0_+(U)$, $t > 0$, *the following estimate holds* :

$$\begin{aligned} \rho\mathcal{M}_{1-\frac{n}{2}}(\phi_\infty) + \rho_t\mathcal{M}_{-t}(g_\infty) &\leq \lambda^+_c(\phi_\infty,g_\infty) \leq \\ &\leq (f_\infty)^{-1} \min_{x_c\in\sum} \left((v(x_0))^{-1}\mathcal{N}_{x_0}(\phi_\infty) + \mathcal{N}_{x_0}(g_\infty)\right) \end{aligned} \tag{1.4.2}$$

where the constants ρ,ρ_t *are given by the formulae* :

$$\rho = (f_\infty)^{-1}\text{meas}(\partial U) \ . \ \min_{x\in U} \mathcal{M}_{\frac{n-2}{n}}(v(x)^{-1}E(x,.))$$

$$\rho_t = (f_\infty)^{-1}\text{meas}(U) \ . \ \min_{x\in U} \mathcal{M}_{\frac{t}{t+1}} (v(x)^{-1}G(x,.)).$$

Proof. The second part of (1.4.2) is obtained by estimating the maximin in (1.3.3) by the value of $\Lambda^+_{\phi_\infty,g_\infty}$ at $x = x_o$.

In order to prove the first part of (1.4.2), note that for positive functions h_1,h_2, and for $p > 1$, $p^{-1} + q^{-1} = 1$, Hölder's inequality can by rewritten as follows :

$$\int h_1 \geq \left(\int h_1^{\frac{1}{p}} h_2\right)^p \left(\int h_2^q\right)^{-\frac{p}{q}}. \qquad (1.4.3)$$

One has :

$$\begin{aligned} f_\infty \lambda_c^+(\phi_\infty,g_\infty) &= \min_{x\in U} ((v(x)^{-1} \int_{\partial U} E(x,y')\phi_\infty(y')d\sigma_{y'} + \\ &= (v(x))^{-1}\int_U G(x,y)g_\infty(y)dy) \\ &\geq \min_{x\in U} ((v(x))^{-1} \int_{\partial U} E(x,y')\phi_\infty(y')d\sigma_{y'}) + \\ &+ \min_{x\in U} ((v(x))^{-1} \int_U G(x,y)g_\infty(y)dy). \end{aligned}$$

In order to estimate the first term from below, one applies (1.4.3) with $p = (n-2)^{-1}n$ and

$$h_1(y') = (v(x))^{-1}E(x,y')\phi_\infty(y'),\ h_2(y') = \phi_\infty(y')^{-\frac{1}{p}}.$$

The second term can be estimated similarly, with $p = t^{-1}(t+1)$ and

$$h_1(y) = (v(x))^{-1}G(x,y)g_\infty(y),\ h_2(y) = g_\infty(y)^{-\frac{1}{p}}.$$

This yields the claim (1.4.2).

For the rest of this section, U is assumed to be the unit ball in $\mathbb{R}^n$. If one has :

$$A_\infty(x,\frac{\partial}{\partial x}) = -\Delta, \quad f_0(s) = \text{sgn } s. \tag{1.4.4}$$

$$g_\infty(x) \equiv 0, \tag{1.4.5}$$

the estimate (1.4.2) takes the following form :

$$2n\, \mathcal{M}_{1-\frac{n}{2}}(\phi_\infty) \leq \lambda_c^+(\phi_\infty,0) \leq 2n\, \mathcal{M}_1(\phi_\infty). \tag{1.4.6}$$

Indeed, in this case, the number ρ can be computed as follows :

$$\rho = \min_{|x|<1} 2n(\Omega_n)^{-1} \int_{|Y'|=1} |x-y'|^{2-n} d\sigma_{y'} = \min_{|x|<1} 2n = 2n \text{ if } n \geq 3$$

$$\rho = \min_{|x|<1} 4(\Omega_2)^{-1} \int_{|Y'|=1} \ln|x-y'|\, d\sigma_{y'} = \min_{|x|<1} 4 = 2n \quad \text{if } n = 2$$

where Ω_n denotes the surface of the unit ball in $\mathbb{R}^n$.

The following result shows that the estimate (1.4.6) is sharp.

<u>Proposition 1.4.2</u>. (i) *If* $n \geq 2$, *then there exist nonconstant function* $\phi_j \in C_+^0(\partial U)$, j=1,2, *such that*

$$\lambda_c^+(\phi_1,0) = 2n\, \mathcal{M}_{1-\frac{n}{2}}(\phi_1), \quad \lambda_c^+(\phi_2,0) = 2n\, \mathcal{M}_1(\phi_2).$$

(ii) *There are no positive constants* ε_n *and* δ_n *such that*

$$2n\, \mathcal{M}_{1-\frac{n}{2}+\varepsilon_n}(\phi) \leq \lambda_c^+(\phi,0), \ \forall\, \phi \in C_+^0(\partial U)$$

or

$$\lambda_c^+(\phi,0) \leq 2n\, \mathcal{M}_{1-\delta_n}(\phi), \quad \forall\, \phi \in C_+^0(\partial U).$$

Proof. For $\xi \in U - \{0\}$ fixed, let $\phi_1(x') = |x'-\xi|^2$, $x' \in \partial U$ (which is, of course, the restriction to ∂U of a linear function).

The corresponding critical solution is the function

$$u_\infty^{2n} = |x-\xi|^2,$$

such that :

$$\lambda_c^+(\phi_1,0) = 2n = 2n\ \mathcal{M}_{1-\frac{n}{2}}(\phi_1). \tag{1.4.7}$$

Let ϕ_2 be the trace on ∂U of the harmonic function $1 + \frac{1}{2} x_1x_2$. The corresponding critical solution is given by the formula

$$u_\infty^{2n}(x) = |x|^2 + \frac{1}{2} x_1x_2$$

such that $\lambda_c^+(\phi_2,0) = 2n = 2n\ \mathcal{M}_1(\phi_2)$, where in the last step the mean value theorem for harmonic functions was used. The claim (ii) is an immediate consequence of (i) and of the following monotonicity property of the mean value : $\mathcal{M}_{t_1}(\phi) < \mathcal{M}_{t_2}(\phi)$ if $t_1 < t_2$ and ϕ is nonconstant on ∂B_1. □ Consider now the one dimensional case $U = (-1,1)$.

Proposition 1.4.3). (i) *Under the assumption (1.4.4), one has :*

$$2\mathcal{M}_{\frac{1}{2}}(\phi_\infty)+2e^{-1}\mathcal{M}_0(g_\infty) \le \lambda_c^+(\phi_\infty,g_\infty) \le 2\mathcal{M}_1(\phi_\infty) + \int_{-1}^{1}(1-|y|)g_\infty(y)dy. \tag{1.4.8}$$

(ii) *Under the assumptions (1.4.4), (1.4.5), one has :*

$$\lambda_c^+(\phi_\infty,0) = 2\mathcal{M}_{\frac{1}{2}}(\phi_\infty). \tag{1.4.9}$$

Proof. A direct computation shows that :

$$\min_{|x|<1} \left((1-x^2)^{-1} \sum_{|x'|=1} (1+x'x)\phi_\infty(x')\right) = 2\mathcal{M}_{\frac{1}{2}}(\phi_\infty). \tag{1.4.10}$$

This proves (1.4.9). The second inequality in (1.4.8) is obtained similarly as the upper bound for λ_0^+ in (1.4.2). In order to prove the first inequality in (1.4.8), one uses (1.4.10) and Hölder's inequality :

$$\begin{aligned}\lambda_c^+(\phi_\infty,g_\infty) &= \min_{|x|<1} 2(1-x^2)^{-1}\Big(\tfrac{1}{2} \sum_{|x'|=1} (1+x'x)\phi_\infty(x') + \\ &+ \int_{-1}^{1} G(x,y)g_\infty(y)dy\Big) \\ &\geq \min_{|x|<1} \left((1-x^2)^{-1} \sum_{|x'|=1} (1+x'x)\phi_\infty(x')\right) + \\ &+ \min_{|x|<1} 2(1-x^2)^{-1} \int_{-1}^{1} G(x,y)g_\infty(y)dy \\ &\geq 2\mathcal{M}_{\frac{1}{2}}(\phi_\infty) + \rho_t\mathcal{M}_t(g_\infty) \qquad \forall t > 0\end{aligned}$$

with ρ_t the same as in Proposition 1.4.1. A computation shows that $\rho_t = 2.(1+(t+1)^{-1}t)^{-t^{-1}(t+1)}$, such that $\lim_{t\downarrow 0} \rho_t = 2e^{-1}$. □

As an extension of (1.4.9) to the multidimensional case onecan mention the following fact : if $\phi_\infty(x')$ can be extended as a linear function on the ball, then $\lambda_c^+(\phi_\infty,0) = \mathcal{M}_{1-n/2}(\phi_\infty)$. (See (1.4.7)). Thus, an explicit formula can be given for $\lambda_c^+(\phi_\infty,0)$ if ϕ_∞ can be extended as a linear function. It seems to be impossible to find such a formula for $\lambda_c^+(\phi_\infty,0)$ for $n > 1$ and $\phi_\infty \in C_+^0(U)$ (see Remark 2 in [9]).

1.5. The critical set in the case of the Laplacian.

It will be assumed that

$$A_\infty(x,\tfrac{\partial}{\partial x}) = -\Delta, \quad f_\infty = 1. \tag{1.5.1}$$

Denote

$$E_c(\phi_\infty, g_\infty) \overset{\text{def}}{=} E_0(v_\infty^{\lambda_c^+(\phi_\infty, g_\infty)}), \forall (\phi_\infty, g_\infty) \in C_+^0(\partial U) \times C^\alpha(U) \tag{1.5.2}$$

where v_∞^λ is the solution of (1.3.1). (If $\lambda_c^+ \geq 0$, then the proof of Theorem 1.3.1 yields : $E_c(\phi_\infty, g_\infty) = E_0(u_\infty^{\lambda_c^+(\phi_\infty, g_\infty)})$).

Theorem 1.5.1. *Let* $(\phi_\infty, g_\infty) \in C_+^0(\partial U) \times C^\alpha(U)$ *with* $\alpha \in (0,1]$. *Then for* $\forall \xi \in E_c(\phi_\infty, g_\infty)$, *there exists* $\psi_\xi \in C_+^0(\partial U)$ *such that :*

$$E_c(\phi_\infty + \delta\psi_\xi, g_\infty) = \{\xi\}, \qquad \delta > 0 \tag{1.5.3}$$

and, moreover, for the solution $v_\delta^{\lambda_c^+}$ *of (1.3.1) corresponding to the data* $\phi_\infty + \delta\psi_\xi, g_\infty$ *the matrix of second derivatives* $D_x^2 v_\delta(\xi)$ *is positive definite,* $\forall \delta > 0$.

Proof. Without loss of generality, it will be assumed that $f_\infty = 1$. Define the function $\psi_\xi \in C_+^0(\partial U)$ as follows :

$$\psi_\xi(x') = |x' - \xi|^2, \quad \forall x' \in \partial U, \quad \xi \in E_c(\phi_\infty, g_\infty), \tag{1.5.4}$$

Denote by w^λ the solution of $\mathcal{T}_\infty^\lambda$ with $g_\infty \equiv 0$ and the boundary condition $\pi_0 w^\lambda = \psi_\xi$. For $\lambda = \lambda_c^+(\psi_\xi, 0)$, one finds easily the corresponding critical solution :

$$w^{\lambda_c^+}(x) = |x - \xi|^2 = \Psi_\xi(x) - 2n \; v(x)$$

where Ψ_ξ is the harmonic function in U, such that $\pi_0 \Psi_\xi = \psi_\xi$ and v is defined by (1.4.1). Hence,

$$\lambda_c^+(\psi_\xi) = \Delta w^{\lambda_c^+}(x) = 2n.$$

Further, since $w^{\lambda_c^+}(x) > 0$, $\forall x \in \bar{U}\setminus\{\xi\}$, one has :

$$\Lambda^+_{\psi_\xi,0}(x) \overset{\text{def}}{=} (v(x))^{-1}\Psi_\xi(x) > 2n = \Lambda^+_{\psi_\xi,0}(\xi) = \min_{y\in\bar{U}} \Lambda^+_{\psi_\xi,0}(y). \quad (1.5.5)$$

Let $u^0_\infty(x)$ denote the solution of the problem $\mathfrak{P}^\lambda$ for $\lambda = 0$. Using (1.5.5) and the fact that ξ is a global minimum of $\Lambda^+_{\phi_\infty,g_\infty}$, one gets :

$$\begin{aligned}\Lambda^+_{\phi_\infty+\delta\psi_\infty,g_\infty}(x) &\overset{\text{def}}{=} (v(x))^{-1}(u^0_\infty(x)+\delta\Psi_\xi(x))\\ &> (v(x))^{-1}u^0_\infty(x)+\delta(v(\xi))^{-1}\Psi_\xi(\xi)\\ &\geq (v(\xi))^{-1}(u^0_\infty(\xi)+\delta\Psi_\xi(\xi)) = \Lambda^+_{\phi_\infty+\delta\psi_\xi,g_\infty}(\xi),\\ &\qquad \forall x \in \bar{U}\setminus\{\xi\}.\end{aligned}$$

Therefore, $\xi \in U$ is the only point where the function $\Lambda^+_{\phi_\infty+\delta\psi_\xi,g_\infty}(x)$ attains its minimum. As a consequence, the function

$$v_1(x) = u^0_\infty(x) - \lambda^+_c(\phi_\infty+\delta\psi_\xi,g_\infty)\, v(x)$$

is such that $E_0(v_1) = \{\xi\}$.

Now the second claim of Theorem 1.5.1 will be proved. A straightforward computation using the relations

$$v_\delta^{\lambda_c}(\xi) = 0, \quad \nabla_x v_\delta^{\lambda_c}(\xi) = 0$$

yields the formula :

$$D^2\Lambda^+_{\phi_\infty+\delta\psi_\xi,g_\infty}(\xi) = (v(\xi))^{-1}D^2v_\delta^{\lambda_c}(\xi), \quad \forall\xi \in E_0(u_\delta^{\lambda_c}). \quad (1.5.6)$$

Hence,

$$D^2 v_\delta^{\lambda_c}(\xi) = v(\xi) D^2 \Lambda^+_{\phi_\infty + \delta\psi_\xi, g_\infty}(\xi)$$

$$= v(\xi)(D^2\Lambda^+_{\phi_\infty, g_\infty}(\xi) + \delta D^2 \Lambda^+_{\psi_\xi, 0}(\xi))$$

$$= v(\xi) D^2 \Lambda^+_{\phi_\infty, g_\infty}(\xi) + \delta D_x^2(|x-\xi|^2)_{|x=\xi}$$

$$\geq \delta\ 2n\ \mathrm{Id}.$$

where Id denotes the identity matrix. □

Now we are going to prove that for $g_\infty \equiv 0$ and for a large class of boundary functions ϕ_∞, the set $E_c(\phi_\infty, 0)$ consists of only one point.

Definition 1.5.2. *For* $\alpha \in (0,1]$, $\mathcal{O}_+^\alpha$ *is defined to be the set of the pairs* $(\phi_\infty, g_\infty) \in C_+^0(\partial U) \times C^\alpha(\bar{U})$, *such that the critical set* $E_c(\phi_\infty, g_\infty)$ *contains only one point* ξ *and such that, for the solution* $v_\infty^{\lambda_c}(\phi_\infty, g_\infty)$ *of (1.3.1), the Hessian* $D^2 v_\infty^{\lambda_c}(\xi)$ *is positive definite at* ξ.

As a consequence of Theorem 1.5.1, the set $\mathcal{O}_+^\alpha$ is dense in $C_+^0(\partial U) \times C^\alpha(\bar{U})$.

Theorem 1.5.3. $^\alpha_+$ *is open in* $C_+^0(\ U) \times C^\alpha(\bar{U})$.

Proof. Let $(\phi_\infty, g_\infty) \in \mathcal{O}_+^\alpha$ and $\{\xi\} = E_c(\phi_\infty, g_\infty)$.
with $\Lambda^+_{\phi_\infty, g_\infty}$ defined by (1.3.3), one gets as in (1.5.6) :

$$D_x^2 \Lambda^+_{\phi_\infty, g_\infty}(\xi) = (v(\xi))^{-1} D_x^2 v_\infty^{\lambda_c}(\xi) \geq 2\gamma\ \ \mathrm{Id} \tag{1.5.8}$$

where $\gamma > 0$. Thus, there exists a constant $\delta_1 > 0$ such that

$$D_x^2 \Lambda^+_{\phi_\infty, g_\infty}(x) \geq \gamma\ \mathrm{Id},\ \forall x \in B_{\delta_1} = \{x \in U\ ||x-\xi| < \delta_1\}. \tag{1.5.9}$$

Since ξ is the only point in the critical set and therefore the only global minimum of $\Lambda^+_{\phi_\infty, g_\infty}(x)$, one can choose a constant $\delta_2 > 0$ such that

$$\Lambda^{+}_{\phi_\infty,g_\infty}(x) \geq \Lambda^{+}_{\phi_\infty,g_\infty}(\xi) + \delta_2, \quad \forall x \in \bar{U}\backslash B_{\delta_1}. \tag{1.5.10}$$

The inequalities (1.5.9), (1.5.10) will be used in order to show that there exists a constant $\rho > 0$ such that for any $(\psi_\infty,h_\infty) \in C^0_+(\partial U) \times C^\alpha(\bar{U})$ satisfying the condition

$$[\psi_\infty-\phi_\infty]_{C^0(\partial U)} + [h_\infty-g_\infty]_{C^\alpha(\bar{U})} \leq \rho,$$

the function $\Lambda^{+}_{\phi_\infty,h_\infty}$ has still only one global minimum. Indeed, one has as a consequence of (1.3.3)

$$\begin{aligned}\lambda^{+}_{c}(\psi_\infty,h_\infty) &= \min_{x\in U} \Lambda^{+}_{\psi_\infty,h_\infty}(x) \leq \Lambda^{+}_{\psi_\infty,h_\infty}(\xi) \leq \\ &\leq \Lambda^{+}_{\phi_\infty,g_\infty}(\xi)+\rho(1+(v(\xi))^{-1}).\end{aligned} \tag{1.5.11}$$

Since

$$\begin{aligned}\Lambda^{+}_{\psi_\infty,h_\infty}(x) &\geq (v(x))^{-1}(\min \psi_\infty+v(x)\min h_\infty) \\ &\geq (v(x))^{-1}(\min \phi_\infty-\rho) + g_\infty - \rho,\end{aligned}$$

and since the function $v(x)$ is zero on the boundary ∂U, one can choose $\delta_3 > 0$ such that for ρ sufficiently small, all global minima of $\Lambda^{+}_{\psi_\infty,h_\infty}$ are contained in the set

$$U_{\delta_3} = \{x \in U \mid \mathrm{dist}(x,\partial U) \geq \delta_3\}.$$

Indeed, the definition of Λ^+ implies :

$$\Lambda^{+}_{\psi_\infty,h_\infty}(x) \geq \Lambda^{+}_{\phi_\infty,g_\infty}(x)-\rho(1+(v(x))^{-1}).$$

Further, (1.5.10), (1.5.11) yield the following inequalities :

$$\Lambda^+_{\psi_\infty,h_\infty}(x) \geq \Lambda^+_{\phi_\infty,g_\infty}(x)-\rho(1+v(x)^{-1})$$

$$\geq \Lambda^+_{\phi_\infty,g_\infty}(\xi)+\delta_2-\rho(1+(v(x))^{-1})$$

$$\geq \Lambda^+_{\phi_\infty,h_\infty}(\xi)+\delta_2-\rho(2+(v(x))^{-1}+(v(\xi))^{-1})$$

$$\geq \min \Lambda^+_{\psi_\infty,h_\infty} + \frac{\delta_2}{2}, \quad \forall x \in U_{\delta_3} \setminus B_{\delta_1},$$

provided that ρ is sufficiently small. Therefore, for such ρ, all global minima of $\Lambda^+_{\psi_\infty,h_\infty}$ are contained in the set B_{δ_1}. The following estimate, however, shows that the functions $\Lambda^+_{\psi_\infty,h_\infty}$ are strictly convex in B_{δ_1} for sufficiently small. Using the interior Schauder estimate and (1.5.9), one finds for $x \in B_{\delta_1}$:

$$D^2_x\Lambda^+_{\psi_\infty,h_\infty}(x) = D^2_x\Lambda^+_{\phi_\infty,g_\infty}(x) +$$

$$+ D^2_x((v(x))^{-1}(\int_{\partial U} E(x,y')(\psi_\infty(y')-\phi_\infty(y'))d\sigma_{y'} +$$

$$+ \int_U G(x,y)(h_\infty(y)-g_\infty(y))dy$$

$$D^2_x\Lambda^+_{\psi_\infty,h_\infty}(x) \geq (\gamma-C\rho)\mathrm{Id}, \quad \forall x \in B_{\delta_1} \tag{1.5.12}$$

where $C > 0$ is some constant.

Hence, $\Lambda^+_{\psi_\infty,h_\infty}$ has for $\rho \ll 1$ only one global minimum and the set $E_c(\psi_\infty,h_\infty)$ contains only one point $\eta = \eta(\psi_\infty,h_\infty) \in B_{\delta_1}$. (1.5.8) and (1.5.12) yield that the matrix of second derivatives of the corresponding critical solution at the point η is positive definite. □

The next result follows immediately from the Teorems 1.5.1, 1.5.3 :

Corollary 1.5.4. *The complement of* $\mathcal{O}_+^\alpha$ *in* $C_+^0(\partial U) \times C^\alpha(\bar{U})$ *is nowhere dense.*

Theorem 1.5.5. *Let* $(\phi_\infty^0, g_\infty^0) \in \mathcal{O}_+^\alpha$. *Then the functional*

$$C_+^0(\partial U) \times C^\alpha(\bar{U}) \ni (\phi_\infty, g_\infty) \to \lambda_c^+(\phi_\infty, g_\infty) \in \mathbb{R} \tag{1.5.13}$$

is Frechet-differentiable at $(\phi_\infty^0, g_\infty^0)$ *and its first variation in the direction* (ψ_∞, h_∞) *is given by the formula :*

$$\delta\lambda_c^+(\phi_\infty^0, g_\infty^0) \circ (\psi_\infty, h_\infty) = (v(\xi))^{-1}\Big(\int_{\partial U} E(\xi,y')\psi_\infty(y')d\sigma_{y'} + \\ + \int_U G(\xi,y)h_\infty(y)dy\Big). \tag{1.5.14}$$

Here $\{\xi\} = E_c(\phi_\infty^0, g_\infty^0)$, G is Green's function of the Dirichelt problem with zero boundary condition for the Poisson equation in U, and the function E and v are defined by $E(x,y') = \pi_0 \frac{\partial}{\partial N_{y'}} G(x,y')$ and (1.4.1), respectively.

Proof. First, one has to show that the function

$$s \to \lambda_c^+(\phi_\infty^0 + s\psi_\infty, g_\infty^0 + sh_\infty) \tag{1.5.15}$$

is differentiable at $s = 0$, $\forall (\psi_\infty, h_\infty) \in C^0(\partial U) \times C^\alpha(\bar{U})$.

Since $\mathcal{O}_+^\alpha$ is open in $C^0(\partial U) \times C^\alpha(\bar{U})$, one has the following formula :

$$\lambda_c^+(\phi_\infty^0 + s\psi_\infty, g_\infty^0 + sh_\infty) = \Lambda^+_{\phi_\infty^0 + s\psi_\infty, g_\infty^0 + sh_\infty}(\xi(s)) \tag{1.5.16}$$

where Λ^+ is defined by (1.3.3) and $\xi(s)$ is well defined by $\{\xi(s)\} = E_c(\phi_\infty^0 + s\psi_\infty, g_\infty^0 + sh_\infty)$ for $|s|$ sufficiently small.

Besides, one has for $\xi(s)$ the following equations :

$$\nabla_x \Lambda^+_{\phi_\infty^0 + s\psi_\infty, g_\infty^0 + sh_\infty}(\xi(s)) = 0.$$

Since, as a consequence of (1.5.8)),

$$D_x^2 \Lambda^+_{\phi^0_\infty, g^0_\infty}(\xi(0)) = v(\xi(0))^{-1} D_x^2 v_\infty^{\lambda_c}(\xi(0)) \geq \gamma \text{ Id}, \; \gamma > 0,$$

the implicit function theorem yields : the function $s \to \xi(s) \in U$ is differentiable for $|s|$ sufficiently small. As a consequence of (1.5.16), the regularity of the function $\Lambda^+(x)$, $x \in U$ and the differentiability of $\xi(s)$, one gets the conclusion that the function (1.5.15) is differentiable at $s = 0$.

A straightforward computation using the relations :

$$v_\infty^{\lambda_c^+(\phi^0_\infty, g^0_\infty)}(\xi(0)) = 0, \quad \nabla_x v_\infty^{\lambda_c^+(\phi^0_\infty, g^0_\infty)}(\xi(0\;)) = 0,$$

then yields the formula (1.5.14). □

<u>Theorem 1.5.6</u>. *Let* $U \subset \mathbf{R}^2$ *be a bounded, simply connected domain and let* $\phi_\infty \in C^0_+(\partial U)$, $g_\infty \equiv 0$. *If the coefficients* $a_{kj}(x)$ *of the differential operator* $A_\infty(x, \frac{\partial}{\partial x})$ *are real analytic in* U, *then the critical set* $E_c(\phi_\infty, 0)$ *consists only of a finite number of points.*

<u>Theorem 1.5.7</u>. *If* $U \subset \mathbf{R}^2$ *is a bounded domain (not necessarily simply connected) and if the coefficients of the differential operator* A_∞ *are real analytic, then* $E_c(\phi_\infty, 0)$ *is the collection of finitely many isolated points and a finite number of closed analytic curves* .

The proof of the Theorems 1.5.6, 1.5.7 stated above is similar to the proof of Theorem 5 and Corollary 2 in [9].

It should be mentioned that even if U is the unit disk in $\mathbf{R}^2$, there exist functions $\phi_\infty \in C^0_+(\partial U)$, such that the set $E_c(\phi_\infty, 0)$ contains more than one point. Moreover, in this case the set $E_0(u_\infty^\lambda)$ is not necessarily connected (see [9]).

1.6. The asymptotic behaviour of the solution to $\mathcal{P}_\infty^\lambda$ when $\lambda \to +\infty$.

In this section, an asymptotic formula for the solution u_∞^λ of $\mathcal{P}_\infty^\lambda$ is indicated and an error estimate in the maximum norm is stated. One uses super- and subsolutions of special type and the maximum principle for establishing this result (see [9], Theorem 7 for the case $A_\infty = -\Delta$). First, consider the case

$$f_{-\infty} < 0 < f_\infty. \tag{1.6.1}$$

For simplicity, it will be assumed from now on, that

$$\phi_\infty \in C_+^0 \cap C^2(\partial U) \tag{1.6.2}$$

For $x \in U$ in a sufficiently small neighbourhood of ∂U, let $x' \in \partial U$ and ρ be defined by

$$|x-x'| = \text{dis}(x,\partial U) = \min_{y'\in\partial U} |x-y'| = \rho.$$

Using the coordinates x',ρ, one can rewrite the operator A_∞ as follows :

$$A_\infty(x,\frac{\partial}{\partial x}) = - a(x',\rho) \frac{\partial^2}{\partial\rho^2} + B$$

Here $a(x',\rho)$ is a smooth positive function and the differential operator B has orders 2 and 1 in $\frac{\partial}{\partial x'}$ and $\frac{\partial}{\partial\rho}$, respectively. Using the notation

$$a(x') = a(x',0) = \sum_{1\le k,j\le n} a_{jk}^\infty(x')N_k(x')N_j(x')$$

and the normalized distance ρ_1,

$$\rho_1 = (\lambda f_\infty.(2a(x'))^{-1})^{1/2}\rho ,$$

one has :

Theorem 1.6.1. *Under the assumptions (1.6.1), (1.6.2), the function*

$$w_\infty^\lambda(x) = ((\phi_\infty(x'))^{1/2}-\rho_1)_+^2, \tag{1.6.3}$$

where $s_+ = \max(s,0)$, *is an asymptotic solution of* $\mathfrak{T}_\infty^\lambda$ *such that*

$$\|u_\infty^\lambda - w_\infty^\lambda\|_{C^0(\bar{U})} \leq C\lambda^{-1/2}, \tag{1.6.4}$$

where the constant C *does not depend on* λ. *Moreover, for the free boundary* $\partial E_0(u_\infty^\lambda)$ *holds :*

$$\partial E_0(u_\infty^\lambda) \subset \{x \in U \,|\, |(2\phi_\infty(x')a(x')(\lambda f_\infty)^{-1})^{1/2}-\rho| \leq C_1\lambda^{-1}\} \tag{1.6.5}$$

where the constant C_1 *does not depend on* λ.

Assuming again (1.6.2), we consider now the case $0 = f_{-\infty} < f_\infty$.

If $g_\infty(x) \geq 0$, $\forall x \in \bar{U}$, then the function (1.6.3) is still an asymptotic solution of $\mathfrak{T}_\infty^\lambda$ when $\lambda \to \infty$. Usint super- and subsolutions, one shows that (1.6.4), (1.6.5) are valid in this case, as well.

If $g_\infty(x) < 0$ $\forall x \in \bar{U}$, $g_\infty \in C^\alpha(\bar{U})$, then it will turn out that for $\lambda \to \infty$, the solution u^λ converges on any $U_1 \subset\subset U$ to the solution $z(x)$ of the linear boundary value problem

$$A_\infty(x,\frac{\partial}{\partial x})z(x) = g\,(x),\ x \in U$$

$$\pi_0 z(x') = 0, \qquad x' \in \partial U.$$

In a neighbourhood of ∂U the solution u_∞^λ has again a boundary layer behaviour as $\lambda \to \infty$.

Let $\beta(x') = \pi_0 \frac{\partial z}{\partial N}(x')$, $x' \in \partial U$, and define $\xi(x')$ to be the positive solution of the quadratic equation $\beta(x')\xi(x') + (\lambda f_\infty/2a(x'))\xi(x')^2 = \phi_\infty(x')$.

Let $U_\xi = \{x \in U \;\, | \rho < \xi(x')\}$ and denote by $w_\infty^\lambda(x)$, $x \in U\backslash U_\xi$, the solution of the following boundary value problem :

$$\begin{cases} A_\infty(x,\frac{\partial}{\partial x})w^\lambda_\infty(x) = g_\infty(x), & x \in U\backslash U_\xi \\ w^\lambda_\infty(x') = 0, & x' \in \partial U_\xi\backslash\partial U \end{cases} \tag{1.6.6}$$

On the set U_ξ, we define w^λ_∞ as follows :

$$w^\lambda_\infty(\rho,x') = \beta_\lambda(x')(\xi(x')-\rho) + (\lambda f_\infty/2a(x'))(\xi(x')-\rho)^2,$$

$$0 < \rho < \xi(x')$$

where $\beta_\lambda(x')$ denotes the restriction of the normal derivative to $\partial U_\xi /\partial U$ of the solution of (1.6.6).
The function w^λ_∞ defined above, is a formal asymptotic solution of $\mathfrak{T}^\lambda_\infty$ when $\lambda \to \infty$:

$$A_\infty(x,\frac{\partial}{\partial x})w^\lambda_\infty + (\lambda+O(\sqrt{\lambda}))f_0(w^\lambda_\infty) + g_\infty(x)\chi_0(w^\lambda_\infty) = g_\infty(x), \quad x \in U$$

$$0 \le g_\infty(x) \le \lambda f_\infty , \qquad x \in \text{Int } E_0(w^\lambda_\infty)$$

$$\pi_0 w^\lambda_\infty(x') = \phi_\infty(x') + O(\lambda^{-1/2}), \qquad x' \in \partial U$$

Note that the boundary condition is satisfied asymptotically because $\xi(x') = O(\lambda^{-1/2})$ when $\lambda \to \infty$ holds uniformly w.r.t. $x' \in \partial U$ and because the Schauder estimate implies that $[\beta-\beta_\lambda]_{C^0(\partial U)} = O(\lambda^{-1/2})$ when $\lambda \to \infty$. Using a maximum principle argument, one finds that w^λ_∞ is an asymptotic solution of $\mathfrak{T}^\lambda_\infty$ when $\lambda \to \infty$.

Note that in the case $f_{-\infty} = 0 < f_\infty$, independently on the sign of $g(x)$, u^λ converges to the solution u of the following problem $\mathfrak{T}_\infty$:

$$\begin{cases} A_\infty(x,\frac{\partial}{\partial x})u = g_\infty(x)(1-\chi_0(u)), & x \in U \\ 0 \le g_\infty(x), & x \in \text{int}(E_0(u)) \\ u(x) \le 0, & x \in U \\ \pi_0 u(x') = 0, & x' \in \partial U \end{cases}$$

In general, $\mathcal{T}_\infty$ is, of course, a free boundary problem.

It can be reformulated as a minimization problem as follows :

$$\min_{\substack{u\in \overset{\circ}{H}_1(U)\\ u\leq 0}} \int_U \left(\frac{1}{2}\sum a_{kj}(x)u_{x_j}u_{x_k} - g_\infty(x)u(x)\right)dx$$

Finally, consider the case $f_{-\infty} < 0 = f_\infty$.

For $\lambda \geq \max_{x\in U}(f_{-\infty}^{-1} g_\infty(x))$, the function u_∞^λ does not depend upon λ and coincides with the solution u of the following problem $\mathcal{T}_\infty$:

$$\begin{cases} A_\infty u(x) = g_\infty(x)(1-\chi_0(u)), & x \in U \\ g_\infty(x) \leq 0, & x \in \mathrm{int}(E_0(u)) \\ u(x) \geq 0 , & x \in U \\ \pi_o u(x') = \phi_\infty(x'), & x' \in \partial U. \end{cases}$$

The equivalent formulation as a minimization problem reads as follows :

$$\min_{\substack{u\in \overset{\circ}{H}_1(U)\\ \pi_0 u=\phi_\infty \\ u\geq 0}} \int_U \left(\frac{1}{2} \sum a_{kj}(x)u_{x_k}u_{x_j} - g_\infty(x)u(x)\right)dx$$

<u>Remark 1.6.2</u>. One can consider a problem with more general non-linearity :

$$\begin{cases} -\Delta u_\infty^\lambda + \lambda q'(u^\lambda)\chi_+(u_\infty^\lambda) = 0, & x \in U \\ \pi_o u_\infty^\lambda(x') = \phi(x'), & x' \in U \end{cases} \tag{1.6.7}$$

where $\phi(x') > 0$, $\forall x' \in \partial U$, $\lambda \gg \phi$ and $q'(s)$ is monotonically increasing on the interval $[0,\phi_M]$ with $\phi_M = \max_{x'\in\partial U} \phi(x')$.

Then the corresponding asymptotic solution $w_\infty^\lambda(x',\rho)$ is defined by the formula :

$$\int_{w_\infty^\lambda(x',\rho)}^{\phi(x')} \frac{ds}{\sqrt{2q(s)}} = \sqrt{\lambda}\rho, \tag{1.6.8}$$

where $q(s)$ is the primitive of $q'(s)$ normalized by the condition $q(0) = 0$. Moreover for the free boundary $\partial E_0(u_\infty^\lambda)$ of u_∞^λ, solution to (1.6.7), holds :

$$\partial E_0(u) \subset \{x \in U \mid |\int_0^{\phi(x')} \frac{ds}{\sqrt{2\rho q(s)}} - \rho| \le C_1\lambda^{-1}\}, \tag{1.6.9}$$

where $C_1 > 0$ is some constant.

In case of a general second order elliptic operator the distance ρ in (1.6.8), (1.6.9) has to be replaced by the normalized distance ρ_1 defined here above.

II. NON-STATIONARY PROBLEM.

2.1. General properties of the operators considered.

Denote

$$B(Q_T) = L^2(Q_T) \times H_1(U) \times H_{3/2,3/4}(\Gamma_T), \quad 0 < T < \infty, \tag{2.1.1}$$

where $H_{s,r}(\Gamma_T)$ is the Sobolev space of all functions $\phi(x',t)$ such that $D_{x'}^\alpha D_t^m \phi \in L^2(\Gamma_T)$, $\forall\, |\alpha| \le s$, $m \le r$ for $s \ge 0$, $r \ge 0$ integer ; if s and r are not necessarily integer , then $H_{s,r}(\Gamma_T)$ is defined in a standard way by using the partition of unity and the Fourier transform. Denote by $\mathfrak{T}_\varepsilon^\lambda$ the operator associated with the initial-boundary value problem (0.6) :

$$\mathfrak{T}_\varepsilon^\lambda : H_{2,1}(Q_T) \to B(Q_T), \quad 0 < T < \infty \tag{2.1.2}$$

Theorem 2.1.1. *For any given $\varepsilon > 0$, the mapping (2.1.2) is a Lipschitz-continuous homeomorphism.*

Theorem 2.1.2. *If* $(g,\psi,\phi) \in C^0(\bar{Q}) \times C^2(\bar{U}) \times C^{2,1}(\Gamma)$ *and the compatibility condition (0.7) is satisfied, then for* $\forall\alpha \in [0,1)$ *uniformly with respect to* $\varepsilon \in (0,\varepsilon_0]$ *holds :*

$$u_\varepsilon^\lambda \in C^{1,\alpha;(1+\alpha)/2}(\bar{Q}).$$

Theorem 2.1.3. *If* $(g,\psi,\phi) \in C^0(\bar{Q}) \times C^2(\bar{U}) \times C^{2,1}(\Gamma)$ *and (0.7) is satisfied, then the reduced problem* $\mathfrak{P}^\lambda$ *has a well-defined (distributional) solution* $u^\lambda \in C^{1,\alpha;(1+\alpha)/2}(\bar{Q})$, $\forall\alpha \in [0,1)$.

Moreover, the set $\{u_\varepsilon^\lambda\}_{0<\varepsilon\le\varepsilon_0} \subset C^{1,\alpha;(1+\alpha)/2}(Q_T)$, *where* u_ε^λ *is the solution of* $\mathfrak{P}_\varepsilon^\lambda$, *has for* $\forall T < \infty$ *as its only condensation point the solution* u^λ *of* $\mathfrak{P}^\lambda$ *when* $\varepsilon \downarrow 0$, *so that*

$$\lim_{\varepsilon\downarrow 0} \|u_\varepsilon^\lambda - u^\lambda\|_{C^{1,\alpha;(1+\alpha)/2}(\bar{Q}_T)} = 0, \ \forall T < \infty, \ \ \forall\alpha \in [0,1).$$

The proof of the Theorems 2.1.1., 2.1.2 and 2.1.3 will be given in a coming authors' publication.

2.2. Convergence as $t \to \infty$.

For simplicity, we assume that

$$A(x,t,\tfrac{\partial}{\partial x}) \equiv A_\infty(x,\tfrac{\partial}{\partial x}), \ \phi(x',t) \equiv \phi_\infty(x'), \ g(x,t) \equiv g_\infty(x). \qquad (2.2.1)$$

Let $\mu > 0$ be the least eigenvalue of A_∞ in U with Dirichlet boundary condition on ∂U.

Theorem 2.2.1. *The stationary solutions* $u_{\varepsilon,\infty}^\lambda$, u_∞^λ *of the problems* $\mathfrak{P}_\varepsilon^\lambda$ *and* $\mathfrak{P}^\lambda$ *respectively, are asymptotically stable in* $L^2(U)$ *as* $t \to \infty$, *and, moreover, the following estimates hold :*

$$\begin{cases} [u_\varepsilon^\lambda(.,t)-u_{\varepsilon,\infty}^\lambda]_{L^2(U)} \le e^{-\mu t}[\psi-u_{\varepsilon,\infty}^\lambda]_{L^2(U)} & \forall t \ge 0, \quad \forall \varepsilon > 0 \\ [u^\lambda(.,t)-u_\infty^\lambda]_{L^2(U)} \le e^{-\mu t}[\psi-u_\infty^\lambda]_{L^2(U)} & \forall t \ge 0. \end{cases} \tag{2.2.2}$$

Proof. The difference $w_\varepsilon^\lambda = u_\varepsilon^\lambda - u_{\varepsilon,\infty}^\lambda$ is the solution of the following problem :

$$\frac{\partial w_\varepsilon^\lambda}{\partial t} + A_\infty(x,\frac{\partial}{\partial x})w_\varepsilon^\lambda + \lambda(f(\varepsilon^{-1}u_\varepsilon^\lambda)-f(\varepsilon^{-1}u_{\varepsilon,\infty}^\lambda)) = 0, \ (x,t) \in Q$$

$$w_\varepsilon^\lambda(x,0) = \psi(x) - u_{\varepsilon,\infty}^\lambda(x), \quad x \in \bar{U}$$

$$\pi_0 w_\varepsilon^\lambda(x',t) = 0, \quad (x',t) \in \Gamma$$

Multiplying the differential equation with w_ε^λ, integrating by parts and using the fact that f is monotonically increasing, one gets the following inequality :

$$\frac{1}{2}\frac{d}{dt}[w_\varepsilon^\lambda(.,t)]^2_{L^2(U)} + (A_\infty w_\varepsilon^\lambda, w_\varepsilon^\lambda) \le 0, \quad \forall t \ge 0 \tag{2.2.3}$$

Now the inequality

$$(A_\infty w,w) \ge \mu[w]^2_{L^2(U)} \qquad \forall\, w \in \overset{\circ}{H}_1(U) \tag{2.2.4}$$

and Gronwall's Lemma yield the first of the inequalities (2.2.2). The second inequality in (2.2.2) follows from the first one and from (2.1.3). □

Theorem 2.2.2. *The following estimates hold under the assumption* (2.2.1) :

$$[(u_\varepsilon^\lambda)_t(.,t)]_{L^2(U)} \le e^{-\mu t}[A_\infty\psi+\lambda f(\varepsilon^{-1}\psi)-g_\infty]_{L^2(U)} \quad \forall t,\varepsilon > 0 \tag{2.2.5}$$

$$(A_\infty(u^\lambda_\varepsilon - u^\lambda_{\varepsilon,\infty}), u^\lambda_\varepsilon - u^\lambda_{\varepsilon,\infty})^{1/2}_{L^2(U)} \le e^{-\mu t}[A_\infty\psi + \lambda f(\varepsilon^{-1}\psi) - g_\infty]^{1/2}_{L^2(U)} \tag{2.2.6}$$

$$.[\psi - u_{\varepsilon,\infty}]^{1/2}_{L^2(U)} \quad \forall t,\varepsilon > 0$$

Proof. Denote $v^\lambda_\varepsilon = (u^\lambda_\varepsilon)_t$, so that v^λ_ε is the solution of the following problem :

$$\begin{cases} \dfrac{\partial v^\lambda_\varepsilon}{\partial t} + A_\infty(x,\dfrac{\partial}{\partial x})v^\lambda_\varepsilon + \varepsilon^{-1}\lambda f'(\varepsilon^{-1}u^\lambda_\varepsilon)v^\lambda_\varepsilon = 0, & (x,t) \in Q \\ v^\lambda_\varepsilon(x,0) = g_\infty - A_\infty\psi - \lambda f(\varepsilon^{-1}\psi), & x \in \bar{U} \\ \pi_0 v^\lambda_\varepsilon(x',t) = 0, & (x',t) \in \Gamma \end{cases} \tag{2.2.7}$$

Since $f'(s) \ge 0$, one gets, using (2.2.7) :

$$\frac{1}{2}\frac{d}{dt}[v^\lambda_\varepsilon(.,t)]^2_{L^2(U)} + (A_\infty v^\lambda_\varepsilon, v^\lambda_\varepsilon) \le 0 \qquad \forall t \ge 0. \tag{2.2.8}$$

Further, (2.2.4) and Gronwall's lemma yield (2.2.5). Using (2.2.3), (2.2.4) and (2.2.5), one obtains :

$$(A_\infty w^\lambda_\varepsilon, w^\lambda_\varepsilon)^{1/2}_{L^2(U)} \le [(u^\lambda_\varepsilon)_t(.,t)]^{1/2}_{L^2(U)} \; [w^\lambda_\varepsilon]^{1/2}_{L^2(U)}$$

$$\le e^{-\mu t}[A_\infty\psi + \lambda f(\varepsilon^{-1}\psi) - g_\infty]^{1/2}_{L^2(U)} \; [\psi - u^\lambda_{\varepsilon,\infty}]^{1/2}_{L^2(U)} \qquad \square$$

2.3. Convergence for $\varepsilon \downarrow 0$.

For simplicity, it is assumed that

$$g \equiv 0, \; \psi \ge 0, \; \phi \ge 0. \tag{2.3.1}$$

Proposition 2.3.1. *Under the asumptions (0.7), (2.3.1), the following estimates hold :*

$$\|u_\varepsilon^\lambda(x,t)-u^\lambda(x,t)\|_{L^2([0,T],H_1(U))} \le C(T\varepsilon)^{1/2} \quad \forall T > 0,$$

$$\|u_\varepsilon^\lambda(x,t)-u^\lambda(x,t)\|_{L^2([0,\infty],L_2(U))} \le C\varepsilon^{1/2}, \tag{2.3.2}$$

where the constant C *depends only upon* λ, meas U, f, *the ellipticity constant of* A *and upon the first eigenvalue of* $(-\Delta)$ *in* U *with Dirichlet boundary conditions on* ∂U.

Proof. The difference $v_\varepsilon^\lambda = u_\varepsilon^\lambda - u^\lambda$ is a solution of the following problem.

$$\begin{cases} (v_\varepsilon^\lambda)_t - A\, v_\varepsilon^\lambda + \lambda(f(\frac{u_\varepsilon^\lambda}{\varepsilon})-f_0(u^\lambda)) = 0, & (x,t) \in Q \\ v_\varepsilon^\lambda(x,0) = 0, & x \in U \\ \pi_o v_\varepsilon^\lambda(x',t) = 0, & x' \in \partial U,\ t \in \mathbb{R}_+ \end{cases} \tag{2.3.3}$$

Multiplying the differential equation with v_ε^λ and integrating by parts, one obtains

$$\frac{1}{2}\frac{d}{dt}[v_\varepsilon^\lambda]^2_{L^2(U)} + \sum_{k,j}\int_U a_{kj}(x,t)(v_\varepsilon^\lambda)_k(v_\varepsilon^\lambda)_j dx + \lambda\int_{E_0(u^\lambda)\cup E_-(u^\lambda)} f(\frac{u_\varepsilon^\lambda}{\varepsilon})v_\varepsilon^\lambda dx =$$

$$= \lambda\int_{E_+(u^\lambda)} (f_0(u^\lambda)-f(\frac{u_\varepsilon^\lambda}{\varepsilon}))v_\varepsilon^\lambda dx$$

As a consequence of the assumption (2.3.1) and of Proposition 2.4.1 below, $E_-(u^\lambda) = \emptyset$ and $v_\varepsilon^\lambda \ge 0$. Thus,

$$\frac{1}{2}[v_\varepsilon^\lambda(.\,,T)]^2_{L^2(U)} + \gamma\int_{Q_T} |\nabla_x v_\varepsilon^\lambda|^2 dxdt \le \lambda\int_{Q_T} (f_\infty - f(\frac{u_\varepsilon^\lambda}{\varepsilon}))v_\varepsilon^\lambda dxdt$$

where $\gamma > 0$ is the ellipticity constant of A.

Thus,

$$\frac{1}{2}\frac{d}{dt}[v_\varepsilon^\lambda]^2_{L^2(U)} + \gamma\mu[v_\varepsilon^\lambda]^2_{L^2(U)} \le \lambda \int_{E_+(u^\lambda)} (f_o(u^\lambda)-f(\frac{u_\varepsilon^\lambda}{\varepsilon}))v_\varepsilon^\lambda dx \quad (2.3.4)$$

where $\gamma > 0$ is the ellipticity constant and $\mu > 0$ is the least eigenvalue of $-\Delta$ with homogeneous Dirichlet boundary condition on ∂U.

The integral in the right hand side of the last inequality can be estimated as follows :

$$\int_U (f_\infty - f(\frac{u_\varepsilon^\lambda}{\varepsilon}))v_\varepsilon^\lambda dx \le C \int_U \frac{\varepsilon}{\varepsilon+u_\varepsilon^\lambda}(u_\varepsilon^\lambda - u^\lambda)dx \le$$

$$\le C\varepsilon \int_U \frac{u_\varepsilon^\lambda}{\varepsilon+u_\varepsilon^\lambda}dx \le C\ \varepsilon \text{ meas } U. \quad (2.3.5)$$

Using (2.3.4), (2.3.5) and Gronwall's lemma, one gets the second of the inequalities (2.3.3).

Integrating (2.3.4) over [0,T] one gets the first of the inequalities (2.3.2). Similarly to section 1.2, an improved estimate for the rate of convergence of u_ε^λ to u^λ can be obtained under the same assumptions upon the free boundary of the solution u^λ. However, the construction of asymptotic solutions in this case is somewhat more tedious than for the stationary problem.

2.4. Nonnegative solutions.

Proposition 2.4.1. *If* $g \ge 0$, $\psi \ge 0$, $\phi \ge 0$, *then* $u_\varepsilon^\lambda \ge 0$, $u^\lambda \ge 0$.

Proof. Assume that the set $E_-(u_\varepsilon^\lambda)$ is nonempty. For $\forall (x,t) \in E_-(u_\varepsilon^\lambda)$, one has :

$$(u_\varepsilon^\lambda)_t + A\ u_\varepsilon^\lambda = g(x,t) - \lambda f(\varepsilon^{-1}u_\varepsilon^\lambda) \ge 0.$$

Moreover, $u_\varepsilon^\lambda(x,t) = 0$ for $\forall (x,t) \in \partial E_-(u_\varepsilon^\lambda)$. The maximum principle yields $u_\varepsilon^\lambda \ge 0$ in $E_-(u_\varepsilon^\lambda)$. Thus one obtains a contradiction. The nonnegativity of u^λ is proved similarly. □

Let v_ε^λ be the solution of the following linear problem :

$$\begin{cases} \dfrac{\partial v_\varepsilon^\lambda}{\partial t} + A(x,t,\dfrac{\partial}{\partial x})v_\varepsilon^\lambda + \dfrac{\lambda}{\varepsilon}f'(0)v_\varepsilon^\lambda = g, & (x,t) \in Q \\ v_\varepsilon^\lambda(x,0) = \psi(x), & x \in \bar{U} \\ \pi_o v_\varepsilon^\lambda(x',t) = \phi(x',t), & (x',t) \in \Gamma. \end{cases} \tag{2.4.1}$$

Proposition 2.4.2. *Assume* $f(s)$ *to be concave and the data to be nonnegative. Then* $u_\varepsilon^\lambda(x,t) \geq v_\varepsilon^\lambda(x,t), \forall(x,t) \in \bar{Q}$, *where* v_ε^λ *is the solution of (2.4.1).*

Proof. The difference $w_\varepsilon^\lambda = u_\varepsilon^\lambda - v_\varepsilon^\lambda$ is a solution of the following problem :

$$(w_\varepsilon^\lambda)_t + Aw_\varepsilon^\lambda + (f(\varepsilon^{-1}u_\varepsilon^\lambda)-f(\varepsilon^{-1}(u_\varepsilon^\lambda-w_\varepsilon^\lambda)) = h_\varepsilon(x,t), \quad (x,t) \in Q$$

$$w_\varepsilon^\lambda(x,0) = 0, \quad x \in U$$

$$\pi_0 w_\varepsilon^\lambda(x',t) = 0, \quad (x',t) \in \Gamma,$$

where the function

$$h_\varepsilon = -f(\varepsilon^{-1}v_\varepsilon) + \varepsilon^{-1}f'(0)v_\varepsilon^\lambda$$

is nonnegative since f is concave. The maximum principle yields : $w_\varepsilon^\lambda \geq 0$ in $\bar{Q}$.

Proposition 2.4.3. *If* $u_\varepsilon^\lambda \geq 0$, *then* u_ε^λ *is a monotonically increasing function of* $\varepsilon > 0$.

Proof. Let $w_\varepsilon^\lambda = \dfrac{\partial u_\varepsilon^\lambda}{\partial \varepsilon}$. Then w_ε^λ is the solution of the problem

$$\begin{cases} \frac{\partial w_\varepsilon^\lambda}{\partial t} + Aw_\varepsilon^\lambda + \varepsilon^{-1}\lambda f'(\varepsilon^{-1}u_\varepsilon^\lambda)w_\varepsilon^\lambda = \varepsilon^{-2}\lambda f'(\varepsilon^{-1}u_\varepsilon^\lambda)u_\varepsilon^\lambda, & (x,t) \in Q, \\ w_\varepsilon^\lambda(x,0) = 0, & x \in \bar{U} \\ \pi_o w_\varepsilon^\lambda(x',t) = 0, & (x',t) \in \Gamma \end{cases}$$

As a consequence of the maximum principle, one finds $w_\varepsilon^\lambda(x,t) \geq 0$, $\forall(x,t) \in \bar{Q}$, since $f'(s) \geq 0$, $u_\varepsilon^\lambda \geq 0$. □

2.5. Special solutions of Cauchy's problem for the reduced operator.

In this section, we assume that

$$A(x,t,\frac{\partial}{\partial x}) = -\Delta$$

$$U = \mathbb{R}^n$$

$$f_0(s) = \text{sgn}s, \ \forall s \in \mathbb{R}\backslash\{0\}, \quad f_0(0) = 0$$

$$f(x,t) = 0 \ \ \forall(x,t) \quad \mathbb{R}^n \times \ \mathbb{R}_+.$$

We indicate the following two types of special nonnegative solutions of Cauchy's problem for the reduced operator.

(i) Travelling waves solutions :

$$u^\lambda(x,t;w,\xi) = \lambda w^{-1}(x.\xi-wt)_+ - \lambda|\xi|^2 w^{-2}[1-\exp(-w|\xi|^{-2}(x.\xi-wt)_+)] \quad (2.5.1)$$

where $\xi \in \mathbb{R}^n$, $w \in \mathbb{R}\backslash\{0\}$ and $s_+ = \max(s,0)$.

(ii) Similarity solutions :

These are solutions of the form

$$u^\lambda(x,t) = t\ v\ (|x|t^{-1/2}), \quad (2.5.2)$$

where $v\ (s)$, $s \in \mathbb{R}_+$, are nonnegative solutions of the following ordinary differential equation

$$-v''(s) - \left(\frac{s}{2}+\frac{n-1}{s}\right)v'(s) + v(s) + \lambda\chi_+(v(s)) = 0, \quad s > 0, \tag{2.5.3}$$

so that $v(s) \geq 0$ is given by the formula

$$v(s) = (s^2+2n)[c_1 \int_1^s \xi^{1-n}(\xi^2+2n)^{-2}e^{-\xi^2/4}d\xi+c_2] - \lambda \tag{2.5.4}$$

If n=1, one finds the solution of the Cauchy problem

$$\begin{aligned} &\frac{\partial u^\lambda}{\partial t} - \frac{\partial^2 u^\lambda}{\partial x^2} + \lambda\chi_+(u^\lambda) = 0, \quad x \in \mathbb{R}, \quad t > 0 \\ &u^\lambda(x,0) = x_+^2 \end{aligned} \tag{2.5.5}$$

where has the form (2.5.2), where

$$v(s) = \{(s^2+2)[a+b \int_0^s (\xi^2+2)^{-2}\exp(-\xi^2/4)d\xi]-\lambda\}\mathcal{H}(s-\alpha) \tag{2.5.6}$$

In this case, the free boundary is a parabola $x = \alpha\sqrt{t}$. One gets a system of three equations for the parameters a,b,α, which after elimination of a and b leads to the following functional equation for the free boundary parameter α :

$$(\alpha^2+2)^{-1} - 2\alpha \exp(\alpha^2/4) \int_\alpha^\infty (\xi^2+2)^{-2}\exp(-\xi^2/4)d\xi = \lambda^{-1}. \tag{2.5.7}$$

For $\forall\, \lambda > 0$ the equation (2.5.7) has a well-defined solution $\alpha \in \mathbb{R}$. For $\lambda=2$, one gets the stationary solution $u(x,t) = x_+^2$.

2.6. Short time asymptotics.

Consider the Cauchy problem :

$$\begin{cases} u_t^\lambda - u_{xx}^\lambda + \lambda\chi_+(u^\lambda) = 0, \quad x \in \mathbb{R}, \quad t > 0 \\ u^\lambda(x,0) = \psi(x) \end{cases} \tag{2.6.1}$$

where $\psi(x) \equiv 0$ for $x < 0$, $\psi(x) = \gamma x_+^2 + O(x_+^3)$ for $x_+ \to 0$ and $\psi(x) \leq Cx^2$,

$\forall x \in \mathbb{R}_+$ with some constant $C > 0$.

Denote by $w^\lambda(x,t)$ the corresponding similarity solution :

$$\begin{cases} w_t^\lambda - w_{xx}^\lambda + \lambda\chi_+(w^\lambda) = 0, & x \in \mathbb{R},\ t > 0 \\ w^\lambda(x,0) = \gamma x_+^2, & \end{cases} \tag{2.6.2}$$

$w^\lambda(x,t)$ being defined by the formula $w^\lambda(x,t) = tv^\lambda(\frac{x}{\sqrt{t}})$ with $v^\lambda(s)$ given by (2.5.6), (2.5.7).

Then the following estimate holds :

$$\sup_{(x,t)\in Q_{a,T}^\delta} |u^\lambda(x,t) - w^\lambda(x,t)| \le C_{a,T}\delta, \quad \forall \delta > 0, \tag{2.6.3}$$

where $Q_{a,T}^\delta = \{(x,t) \mid 0 \le t \le T\delta^2, |x| \le a\delta\}$ and the constant $C_{a,T}$ does not depend of δ.

One proves (2.6.3) using the method introduced in [21].

2.7. Asymptotics for $\lambda \to \infty$.

In order to avoid unnecessary technical complications we consider here the case of one space variable $x \in U = (-1,1)$ and of special (constant) initial and boundary conditions. Namely, consider the problem $\mathcal{P}_\varepsilon^\lambda$:

$$\begin{cases} u_t^\lambda - u_{xx}^\lambda + \lambda\chi_+(u^\lambda) = 0, & (x,t) \in U \times \mathbb{R}_+ \\ u^\lambda(x,0) = 1. & x \in U \\ \pi_0 u^\lambda(x',t) = 1, & (x',t) \in \partial U \times \mathbb{R}_+ \end{cases} \tag{2.7.1}$$

Let $v(x,t)$ be the solution of the problem :

$$\begin{cases} v_t - v_{xx} = 1, & (x,t) \in U \times \mathbb{R}_+ \\ v(x,0) = 0, & x \in U \\ \pi_0 v(x',t) = 0, & (x',t) \in \partial U \times \mathbb{R}_+, \end{cases} \tag{2.7.2}$$

and let $t_c(\lambda)$ be defined by the relation :

$$v(0,t_c(\lambda)) = \lambda^{-1}. \qquad (2.7.3)$$

Then for $t \in [0,t_c(\lambda))$ holds :

$$u^\lambda(x,t) = 1-\lambda v(x,t).$$

Denote

$$\gamma = \gamma(\lambda) = \lambda(1-v(0,t_c(\lambda)) \qquad (2.7.4)$$

and let $w_\gamma^\lambda(x,t)$ be the similarity solution of the Cauchy problem :

$$\begin{cases} (w_\gamma^\lambda)_t - (w_\gamma^\lambda)_{xx} + \lambda\chi_+(w_\gamma^\lambda) = 0, & (x,t) \in \mathbb{R} \times \mathbb{R}_+ \\ w_\gamma^\lambda(x,0) = \gamma x^2 & \end{cases} \qquad (2.7.5)$$

<u>Theorem 2.7.1</u>. *For any constants* T > 0, a > 0, *there exists a constant* C = C(T,a) *such that for the solution* $u^\lambda(x,t)$ *the following inequality holds:*

$$\sup_{\substack{0\le t-t_c(\lambda)\le T\lambda^{-1} \\ |x|\le a\lambda^{-1/2}}} |u^\lambda(x,t)-w_\gamma^\lambda(x,t-t_c(\lambda))| \le C(T,a)\lambda^{-1}, \qquad (2.7.6)$$

where $t_c(\lambda)$, γ, $w_\gamma^\lambda(x,t)$ *are defined by* (2.7.2)-(2.7.5).

<u>Remark 2.7.2</u>. One finds easily that $t_c(\lambda) = \lambda^{-1} + o(\lambda^{-1})$. Furthermore, for $0 \le t < t_c(\lambda)$ the following asymptotic formula holds :

$$u^\lambda(x,t) = (1-\lambda t) + O(\lambda^{-1}) \text{ for } \lambda \to \infty. \qquad (2.7.7)$$

while for $\sqrt{\lambda}t \gg 1$ one has :

$$u^\lambda(x,t) \sim u_\infty^\lambda(x), \ \lambda \to \infty \ , \qquad (2.7.8)$$

where

$$u_\infty^\lambda(x) = (\lambda/2)(|x|-1+\sqrt{2\lambda^{-1}})_+^2$$

is the solution of the corresponding stationary problem.

For $t_c(\lambda) < t < c\lambda^{-1/2}$ one has :

$$u^\lambda(x,t) \sim v(\lambda(t-t_c(\lambda)), \ \sqrt{\lambda}\,\text{dist}(x,\partial U)), \quad \lambda \to \infty, \tag{2.7.9}$$

where $v(\tau,\zeta)$ is the solution of the problem :

$$\begin{cases} v_\tau - v_{\zeta\zeta} + \chi_+(v) = 0, & (\zeta,\tau) \in \mathbb{R}_+ \times \mathbb{R}_+ \\ v(\zeta,0) = 1, & \zeta \in \mathbb{R}_+ \\ v(0,\tau) = 1, & \tau \in \mathbb{R}_+ \end{cases} \tag{2.7.10}$$

Introducing $w = v_\tau$ one can reformulate (2.7.10) as the following slightly modified Stefan problem :

$$\begin{aligned} & w_\tau - w_{\zeta\zeta} = 0, \qquad 0 < \zeta < s(\tau),\ \tau \in \mathbb{R}_+ \\ & w(0,\tau) = 0, \qquad \tau \in \mathbb{R}_+ \\ & w(\zeta,0) = -1, \qquad \zeta \in \mathbb{R}_+ \\ & w(s(\tau),\tau) = 0, \quad w_x(s(\tau),\tau) + \dot{s}(\tau) = 0, \quad \tau \in \mathbb{R}_+ \\ & s(\infty) = \sqrt{2} \end{aligned} \qquad (2.7.11$$

the curve $\zeta = s(\tau)$ being the free boundary for the solution $w(\zeta,\tau)$ of (2.7.11).

One has also :

$$\lim_{\tau\to\infty} v(\zeta,\tau) = (1-\zeta/\sqrt{2})_+^2. \tag{2.7.12}$$

The proof of Theorem 2.7.1. and the claims stated in Remark 2.7.2, as well as the corresponding generalizations, will be presented elsewhere.

This research has been partially supported by the European Research Office of the US Army under the contract N° DAJA-37-82-C-0731.

REFERENCES.

[1] C. Baiocchi, Su un problema a frontiera libera conesso a questioni di idraulica, Ann. Mat. Pura ed Appl. (4), 92 (1972), pp. 107-127.

[2] C.M. Brauner, B. Nicolaenko, Singular perturbations and free boundary problems, in : Computing methods in Applied Sciences and Engineering, R. Glowinsky and J.-L. Lions, eds., North-Holland, 1980.

[3] H. Brézis, Opérateurs maximaux monotones, North-Holland, 1973.

[4] H. Brézis, D. Kinderlehrer, The smoothness of solutions to nonlinear variational inequalities, Indiana Math. Journ., 23 (1974), pp. 831-844.

[5] I. Ekeland, R. Témam, Convex analysis and variational problems, North-Holland, 1976.

[6] L.S. Frank, E.W.C. van Groesen, Singular Perturbations of an Elliptic operator with Discontinuous Nonlinearity, in : Analytical and Numerical Approaches to Asymptotic Problems in Analysis, O. Axelsson, L.S. Frank, A. van der Sluis (eds.,) North-Holland, 1981.

[7] L.S. Frank, W.D. Wendt, On a singular perturbation in the kinetic theory of enzymes, in : Theory and applications of singular perturbations, W. Eckhaus, E.M. de Jager (eds.), Springer, 1982.

[8] L.S. Frank, W.D. Wendt, Sur une perturbation singulière en théorie des enzymes, C.R.A.S. Paris, t. 294 (1982), pp. 741-744, Solutions asymptotiques pour une classe de perturbations singulières elliptiques semilinéaires, C.R.A.S. Paris, t. 295 (1982), pp. 451-454, Sur une perturbation singulière parabolique en théorie cinétique des enzymes, C.R.A.S. Paris, t. 295 (1982),

pp. 731-734.

[9] L.S. Frank, W.D. Wendt, On an elliptic operator with discontinuous nonlinearity, J. Diff. Eq., 1984.

[10] Free boundary problems I, II, Proceedings of a seminar at Pavia, E. Magenes (ed.), Rome, 1980.

[11] A. Friedman, Variational principles and free boundary problems, Pure and Applied Mathematics, Wiley Interscience Series, 1982.

[12] R. Goldman, O. Kedem, E. Katchalski, Papain - Collodoin Membranes II. Biochemistry 10 (1971), pp. 165-172.

[13] D. Kinderlehrer, L. Nirenberg, The smoothness of the free boundary in the one phase Stefan problem, Séminaire Collège de France, 1976.

[14] H. Lewy, G. Stampacchia, On the regularity of the solution of a variational inequality. Comm. Pure Appl. Math. 22 (1969), pp. 153-188.

[15] J.-L. Lions, G. Stampacchia, Variational inequalities, Comm. Pure Appl. Math. 20 (1967), pp. 493-519.

[16] J.-L. Lions, Perturbations singulières dans les problèmes aux limites et en contrôle optimale, Springer, 1973.

[17] R. Témam, Remarks on a free boundary problem arising in plasma physics. Comm. PDE 2 (1977), pp. 563)585.

[18] A. Trenogin, Development and applications of Vishik-Lyusternik's asymptotic method, Uspekhi Math. Nauk. 25 (1970), pp. 123-156.

[19] M. Vishik, L. Lyusternik, Regular degeneration and boundary layer for linear differential equations with small parameter, Uspekhi Mat. Nauk 12 (1957), pp. 3-122, AMS Transl. (2), 20 (1962), pp. 239-364.

[20] M. Vishik, L. Lyusternik, The asymptotic behaviour of solutions of linear differential equations with large or quickly varying coefficients and conditions at the boundary, Uspekhi Mat. Nauk, 15 (1960), pp. 27-95.

[21] S. Kamin, The asymptotic behavior of the solution of the filtration equation, Israel J. Math., 14 (1973), pp. 76-87.

L.S. FRANK
Department of Mathematics
University of Nijmegen
Toernooveld
6525 ED NIJMEGEN
THE NETHERLANDS

W.D. WENDT
Department of Applied Mathematics
University of Bonn
Wegelerstr. 10
D-53 BONN 1
GERMANY

M GARONNI

On bilateral evolution problems of non-variational type

The purpose of this lecture is to present some results about bilateral problems related to a second order linear or non-linear elliptic or parabolic operator with the principal part not in divergence form.

The results we consider here have been obtained by M.A. Vivaldi and myself (see [7], [8] and [9]).

Problems of this kind arise for instance in the control theory of stochastic games when the evolution of the involved system is governed by a diffusion with continuous coefficients. From this point of view these problems were introduced by A. Friedman in [5] and developed by A. Bensoussan and J.L. Lions in [1].

Obviously these problems cannot be formulated as variational inequalities. This entails a lot of difficulties ; there is no general existence or uniqueness result as in the variational context and one can prove that a "strong" solution exists only if the obstacles are "very regular".

On the other hand it is important to introduce a generalized solution which is an appropriate substitute of solution when the latter does not exist. This happens for instance if the obstacles are merely continuous. Similarly it is not even possible to approximate directly the strong solution by using variational methods based on the finite elements method.

In this lecture we will dwell upon the following approximation results : we consider the "regularizing" (variational) problems and we estimate in the uniform norm the order of convergence of the solutions in terms of the order of convergence of the corresponding coefficients.

We sketch the proof in the simplest case, that of strong solutions in the linear case and we refer to [8] and [9] for the proof for generalized solutions in the linear case and for strong solutions in the non linear case, respectively.

Thus, preliminarily, we will only give the essential information about existence, uniqueness and regularity results (see §1, §2, theorems 2.1, 2.3 and 2.4) and we refer to [7] and [8] for the proofs, the details and for the stochastic characterization.

1. PRELIMINARIES AND POSITION OF THE LINEAR PROBLEM.

G is a bounded open subset of $\mathbb{R}^N_x$, with boundary Γ of class C^2. Q is the cylinder $G \times]0,T[$ of $\mathbb{R}^N_x \times \mathbb{R}_t$ and Σ is its lateral boundary $\Gamma \times]0,T[$, where $0 < T < +\infty$.

We set

$$\left\{ \begin{aligned} Lv &= -\sum_{i,j=1}^{N} a_{ij} \frac{\partial^2 v}{\partial x_i \partial x_j} + \sum_{i=1}^{N} a_i \frac{\partial v}{\partial x_i} + a_o v \\ Ev &= \frac{\partial v}{\partial t} + Lv, \end{aligned} \right. \tag{1.1}$$

assuming that

$$\left\{ \begin{aligned} &i) \quad a_{ij} \in C^o(\bar{Q}), \quad \forall i,j = 1,\dots,N \\ &ii) \quad a_i \in L_\infty(Q), \quad \forall i = 0,\dots,N \end{aligned} \right. \tag{1.2}$$

$$\exists k_o > 0 : \sum_{i,j=1}^{N} a_{ij}\xi_i\xi_j \geq k_o|\xi|^2 \text{ on } \bar{Q}, \quad \forall \xi \in R^N. \tag{1.3}$$

Remark 1.1. Of course we can always suppose :

$$\bar{a}_o \equiv \text{ess inf } a_o > 0 \tag{1.4}$$

because we arrive at the condition (1.4) by the usual change of functions.

Note that under the hypothesis $(1.2)_i$ operators like E can not be written in the divergence form. Therefore bilateral (or unilateral problems) for E cannot be formulated as variational inequalities. A correct formulation is : to find $u = u(x,t)$ such that :

$$\begin{cases} u \in W_q^{2,1}(Q),\ q \in (1,+\infty),\quad \psi_1 \le u \le \psi_2 \\ (Eu-f)(u-\psi_1) \le 0 \text{ a.e. in } Q \\ (Eu-f)(u-\psi_2) \le 0 \text{ a.e. in } Q \\ u = 0 \text{ on } \Sigma,\ u(.,0) = 0 \text{ on } G. \end{cases} \tag{1.5}$$

Of course from (1.5) it follows that $(u-\psi_1)(u-\psi_2)(Eu-f) = 0$, a.e. in Q. Our general assumptions about the given functions ψ_i, f are :

$$f \in L_q(Q) \tag{1.6}$$

$$\begin{aligned} &\psi_i \in L_q(Q),\ \psi_1 \le \psi_2 \text{ a.e. in } Q \\ &K_{\psi_1}^{\psi_2} = \{v \in L_q(Q),\ \psi_1 \le v \le \psi_2, \text{ a.e. in } Q\}. \end{aligned} \tag{1.7}$$

We shall denote any solution of (1.5) by $\rho(\psi_i,f,E,W_q^{2,1})$ and we shall call it a "strong" solution.

We only deal with homogeneous Cauchy-Dirichlet condition since the non-homogeneous case can be reduced to the homogeneous one by means of translations.

Observe that for problem (1.5) there is no general existence or uniqueness result, neither in analogous elliptic problems, the basic existence and uniqueness result of G. Stampacchia [13] and of J.L. Lions and of G. Stampacchia [11] cannot be invoked. The first existence and uniqueness results for elliptic or parabolic operators were obtained in [5] and in [1] for "very regular" obstacles through a probabilistic approach.

We now observe that the stochastic context itself suggests the importance of studying the case in which the obstacles are much less regular. Actually, the payoff function $J^{x,t}$ (and therefore the stochastic game) connected with the problem (1.5) is well defined under much weaker assumptions than those in [1] and in [5].

We look for a more general approach to problem P by defining a generalized solution. In this way using analytic tools we obtain the existence and the uniqueness of the strong solution (i.e., belonging to $W_q^{2,1}(Q)$), but under weaker hypotheses than those in [1] and [5].

At the same time we show that a generalized solution exists, is unique and provides an analytic characterization of the "value" of the game. This one is an appropriate substitute for the solution where the latter does not exists. This happens for instance in the case of merely continuous obstacles.

We first state a few results about the unilateral problem which was studied by G.M. Troianiello in [15].

Let

$$K_\psi = \{v \in L_q(Q),\ v \geq \psi,\ \text{a.e. in } Q\} \tag{1.8}$$

where

$$\psi \in L_q(Q).$$

The problem is : to find u such that

$$\begin{cases} u \in W^{2,1}(Q),\ q \in (1,+\infty) \\ u \geq \psi,\ Eu \geq f,\ (Eu-f)(u-\psi) = 0,\ \text{a.e. in } Q \\ u(.,0) = 0 \text{ on } G,\ u = 0 \text{ on } \sum \end{cases} \tag{1.9}$$

We call *uppersolution* of (1.9) any function v such that :

$$\begin{cases} v \in W_q^{2,1}(Q) \\ v \geq \psi,\ Ev \geq f,\ \text{a.e. in } Q \\ v(.,0) \geq 0 \text{ on } G,\ v \geq 0 \text{ on } \sum \end{cases}$$

Let $\sum_E(\psi,f)$ be the set of all uppersolutions of (1.9) that we suppose non empty ; we denote

$$\tau_E(\psi,f) = \inf_{L_q(Q)} \sum_E(\psi,f), \tag{1.10}$$

of course $\tau_E(\psi,f)$ belongs to $L_\infty(Q)$ if ψ belongs to $L_\infty(Q)$, and $q > \frac{N+2}{2}$(e.g.).

The connection between problem (1.9) and the function $\tau_E(\psi,f)$ is clear from proposition 1.1. below, that can easily be proved by extending the proof given in [14] for q = 2 to the case of arbitrary q.

Proposition 1.1. *Let* (1.6) *and* (1.8) *hold. If there exists a function* $u \in W_q^{2,1}(Q)$ *satisfying* (1.9) *then :* $\tau_E(\psi,f) = u$.

In the following we call $\tau_E(\psi,f)$ a *generalized* solution of (1.9).

Of course a formulation such as (1.10) doesn't work for bilateral problems. So we introduce the following implicit system of inequalities connected to problem (1.5).

We follow an approach used by Nakoulima [12] in the case of coercive operators with principal part of divergence form. The system we are interested in is : to find (u_1,u_2) such that

$$\begin{cases} u_i \in W_q^{2,1}(Q),\ i = 1,2,\ q \in (1,+\infty), \\ Eu_1 \geq f_1, u_1 \geq \psi_1 + u_2, \text{ a.e. in } Q, \\ Eu_2 \geq f_2, u_2 \geq -\psi_2 + u_1, \text{ a.e. in } Q, \\ (Eu_1-f_1)(u_1-\psi_1-u_2) = 0 = (Eu_2-f_2)(u_2+\psi_2-u_1), \text{ a.e. in } Q \\ u_1 = 0 = u_2 \text{ on } \sum,\ u_1(.,0) = 0 = u_2(.,0) \text{ on } G. \end{cases} \tag{1.11}$$

We call (u_1,u_2) a *strong* solution of (1.11).

The connection between problems (1.5) and (1.11) is given by the proposition 1.2 that can be proved as in [2].

Proposition 1.2. *If* (u_1,u_2) *is a solution of* (1.11) *then :*

$$u = u_1 - u_2 = \rho(\psi_i, f_1 - f_2, E, W_q^{2,1}(Q)).$$

As a consequence of proposition 1.2 we can derive a solution of problem (1.5) from any solution of the system (1.11). Thus in order to obtain an appropriate substitute for a solution of (1.5) when the strong solution doesn't exist - that is for instance if the obstacles are merely continuous - we define a generalized solution of the system (1.11).

Definition 1. *A genralized solution of* (1.11) *is a pair* $(u_1,u_2) \in (L_q(Q))^2$ *such that :*

$$\begin{cases} u_1 = \tau_E(\psi_1 + u_2, f_1) \\ \\ u_2 = \tau_E(-\psi_2 + u_1, f_2). \end{cases} \tag{1.12}$$

where $\tau_E(.,.)$ *is defined in* (1.10).

Definition 2. *A generalized solution of* (1.5) *is any* $u \in L_q(Q)$ *such that* $u = u_1 - u_2$, *where* (u_1,u_2) *is a generalized solution of* (1.11).

Definition 3. *A generalized solution of* (1.5) *is called regular if* (u_1,u_2) *solution of* (1.12), *belongs to* $(W_q^{2,1}(Q))^2$.

We observe that, since the obstacles in (1.12) are not determined, it is possible that either no solution exists or that several solutions exist.

We refer to [7] and [8] for existence, uniqueness, Hölder continuity and stochastic properties of strong and generalized solutions. Rather we will dwell upon the approximation problems and we will simply sketch the proof for strong solutions in the linear case.

The approximation results for generalized solutions of linear case, as well as for strong solutions of quasi-linear problems will be mentioned.

2. APPROXIMATION RESULTS.

To begin with, it is necessary to mention at least the existence result for strong solution of problem (1.5). Assume

$$\begin{cases} \psi_1 = \bigvee_{j=1}^{r} \psi_1^i, \psi_2 = \bigwedge_{j=1}^{s} \psi_2^j; \psi_i^j \in W_q^{2,1}(Q), \; i = 1,2, \; \psi_1 \leq \psi_2 \text{ on } Q, \\ \psi_1|_\Sigma \leq 0 \leq \psi_2|_\Sigma; \psi_1(.,0) \leq 0 \leq \psi_2(.,0) \text{ on } G. \end{cases} \tag{2.1}$$

The following theorem provides existence, uniqueness of the strong solution and connection between strong and generalized solutions of (1.5).

Theorem 2.1. *Assume* (1.6) *and* (2.1). *Then :*

the set of generalized solutions of (1.5) *is non-empty and there exists a unique regular generalized solution* u

and we have :

$$u = \rho(\psi_i, f, E, W_q^{2,1}(Q)) \tag{2.2}$$

$$\bigwedge_{j=1}^{s} E\psi_2^j \wedge f \leq Eu < f \vee \bigvee_{j=1}^{r} E\psi_1^j. \tag{2.3}$$

We now consider the operators E^n whose coefficients converge to the coefficients of E,

$$E^n = \frac{\partial}{\partial t} + L^n, \; L^n = - \sum_{i,j=1}^{N} a_{ij}^n \frac{\partial^2}{\partial x_i \partial x_j} + \sum_{i=1}^{N} a_i \frac{\partial}{\partial x_i} + a_0$$

such that

$$\begin{cases} a_{ij}^n \in C^\infty(\bar{Q}), \; a_{ij}^n \to a_{ij} \text{ in } C^\circ(\bar{Q}), \text{ as } n \to +\infty \\ a_{ij}^n \text{ satisfying (1.3).} \end{cases} \tag{2.4}$$

We denote by $(1.5)_n$ problem (1.5) with E replaced by E^n. By the regularity assumptions on the a^n_{ij} we can write the operators L^n in divergence form. We give two approximation results. In the first one we consider "regular obstacles" and we approximate the strong solution of (1.5) by the strong solutions of the "variational" problems $(1.5)_n$. In the second one we consider "Hölder continuous obstacles" and we approximate the generalized solution of (1.5) by the generalized solutions of the "variational" problems $(1.5)_n$. We denote by u^n and u the strong as well as the generalized solutions of the bilateral problems $(1.5)_n$ and (1.5), respectively. We want to estimate in the uniform norm the order of convergence of u^n to u in terms of the order of convergence of a^n_{ij} to a_{ij}.

We observe that in the equations the solutions have the same order of convergence as the coefficients. It is sufficient to use $E^n u^n = f = Eu$: in fact one has, for $q > \frac{N+2}{2}$,

$$\|u^n-u\|_{L^\infty} \leq s(q)\|E^n(u^n-u)\|_{L^q} = s(q)\|(E^n-E)u\|_{L^q} \leq$$

$$\leq \; s(q)\|a^n_{ij}-a_{ij}\|_{L^\infty}\cdot\|u\|_{W^{2,1}_q} \leq c(q)s(q)\|f\|_{L^q}\cdot\|a^n_{ij}-a_{ij}\|_{L^\infty} \quad ,$$

For the inequalities, on the contrary, the approximation results cannot be obtained directly from the previous ones by using the Lewy-Stampacchia inequality (2.3). In fact we only have

$$\bigwedge_{j=1}^{s} E^n\psi_2^j \wedge f \leq E^n u^n \leq f \vee \bigvee_{j=1}^{r} E^n\psi_1^j$$

$$\bigwedge_{j=1}^{s} E\psi_2^j \wedge f \leq Eu \leq f \vee \bigvee_{j=1}^{r} E\psi_1^j$$

Yet one can obtain analogous results also for the inequalites. That is :

Theorem 2.2. *Assume*

$$q \in ((N+2) \vee \frac{(N+3).(1-\beta)2}{\delta}, +\infty), \ \delta \in (0,1], \ \beta \in [0,1], \tag{2.5}$$

(1.6) *and* (2.1) *with* q *as in* (2.5), *and* E^n *as in* (2.4) *s.t.* :

$$\begin{cases} \text{i)} \ \|a^n_{ij} - a_{ij}\|_{L_\infty(Q)} \le cn^{-\delta} \\ \\ \text{ii)} \ \|a^n_{ij}\|_{L_\infty(0,T;W^1_\infty(G))} \le cn^{1-\beta} \end{cases} \tag{2.6}$$

Then we have :

$$\|u^n - u\|_{L_\infty(Q)} \le \frac{M(q)}{\bar{a}_0} n^{-\delta+\varepsilon} \tag{2.7}$$

where

$$\varepsilon > \frac{(N+3).(\delta+2-2\beta)}{N+3+q}$$

and

$$M(q) = C\{\|f\|_{L_q(Q)} + \bigvee_{j=1}^{r} \|E\psi_1^j\|_{L_q(Q)} + \bigvee_{j=1}^{s} \|E\psi_2^j\|_{L_q(Q)}\}, \tag{2.8}$$

C *depends only on* q, *on the coefficients of* E, *on the constants in* (2.6) *and on the dimension* N.

We observe that obviously L^n are not coercive. Of course we can make them coercive, but not uniformly because the derivates $\frac{\partial a^n_{ij}}{\partial x_i}$ "explode" (the a^n_{ij} converge to the a_{ij} that aren't lipschitz-continuous). Then hypothesis (2.6) ii) means that this explosion is controlled by $cn^{1-\beta}$.

We give an example of operators E^n satisfying (2.6).

Lemma 2.1. *Assume :*

$$a_{ij} \in L_\infty(0,T;C^{0,\delta}(\bar{G})), \text{ with } \delta \in (0,1] \; ; \tag{2.9}$$

then there exists a sequence of operators E^n *satisfying* (2.4) *and such that :*

$$\|a^n_{ij}-a_{ij}\|_{L_\infty(Q)} \leq cn^{-\delta}\|a_{ij}\|_{L_\infty(0,T;C^{0,\delta}(\bar{G}))} \tag{2.10}$$

$$\|a^n_{ij}\|_{L_\infty(0,T;W^1_\infty(G))} \leq cn^{1-\delta}\|a_{ij}\|_{L_\infty(0,T;C^{0,\delta}(\bar{G}))} \tag{2.11}$$

We only want to give some idea of the kinds of techniques we have used, see [9] for complete proof .

Clearly we can write problem (1.5) in the following form :

$$\int_Q Eu(u-w)dxdt \leq \int_Q f(u-w)dxdt \; ; \quad \forall w \in L_2(Q), \; \psi_1 \leq w \leq \psi_2 \tag{2.12}$$

and problem $(1.5)_n$

$$\begin{cases} \int_Q (E^nu^n+\lambda^nu^n)(u^n-v)dxdt \leq \int_Q (f+\lambda^nu^n)(u^n-v)dxdt \\ \\ \forall v \in L_2(Q), \quad \psi_1 \leq v \leq \psi_2. \end{cases} \tag{2.13}$$

We now choose λ^n large enough to make $E^n + \lambda^n$ coercive on $L_2(0,T;\overset{\circ}{W}{}^1_2(\Omega))$; by $(2.6)_{ii}$ we have :

$$0 \leq \lambda^n \leq cn^{2(1-\beta)}. \tag{2.14}$$

Let us consider

$$d_n \equiv \frac{\lambda^n}{\lambda^n+\bar{a}_0}\|u^n-u\|_{L_\infty(Q)} + \frac{M(q)}{\lambda^n+\bar{a}_0} \cdot \frac{n^{-\delta+\varepsilon}}{3+\omega(N)} \tag{2.15}$$

where $\omega(N) = (\text{meas}_N B_1)^{-1/N}$, B_1 is the ball of $\mathbb{R}^N$ of radius 1. If we choose in (2.12) and (2.13) $w = (u^n-d_n) \vee u$ and $v = (u+d_n) \wedge u^n$ respectively, we obtain

$$\int_Q (E^n+\lambda^n(u^n-u-d_n)(u^n-u-d_n)^+ dxdt \leqslant$$

$$\leqslant \int_Q (\lambda^n(u^n-u)-d_n(a_o+\lambda^n)-(E^n-E)u)(u^n-u-d_n)^+ dxdt.$$

It follows that

$$c\|(u^n-u-d_n)^+\|^2_{L_2(0,T;\overset{\circ}{W}{}^1_2(\Omega))} \leq B_n \tag{2.16}$$

where

$$B_n = \int_Q (\lambda^n(u^n-u)-d_n(a_o+\lambda^n)-(E^n-E)u)(u^n-u-d_n)^+ dxdt. \tag{2.17}$$

A series of computations, which we omit, shows that (2.17) yields

$$B_n \leq \frac{M(q)n^{-\delta}}{3+\omega(N)} \|(u^n-u-d_n)^+\|_{L_1(Q)} \{-n^{\varepsilon}+(u^n-u-d_n)^+\|^{-1/q}_{L_1(Q)}\} \tag{2.18}$$

Obviously this case is not similar to the standard one ; in fact B_n can be zero only if the solution u belongs to $W^{2,1}_{\infty}$. But regularity results of this kind under the above hypotheses are unknown to us. Then we consider the set :

$$F_n = \{(x,t) \in \bar{Q},\ u^n-u-d_n \geq \frac{n^{-\delta+\varepsilon}}{\lambda^n+\bar{a}_o} \cdot \frac{M(q)}{3+\omega(N)}\} \ ; \tag{2.19}$$

we have

$$\text{meas } F_n < \left(\frac{n^{-\delta+\varepsilon}}{\lambda^n+\bar{a}_o}\right)^{N+2} . \tag{2.20}$$

In fact if

$$\text{meas } F_n \geq \left(\frac{n^{-\delta+\varepsilon}}{\lambda^n+\bar{a}_o}\right)^{N+2} \tag{2.21}$$

from (2.18), (2.21), (2.19) and (2.15), by choosing $\varepsilon > \frac{(N+3)(2-2\beta+\delta)}{q+N+3}$ we obtain $B_n \le 0$ definitely and hence by (2.16) : $u^n - u \le d_n$; which contradicts (2.21). Thus (2.20) holds.

Since $u^n - u$ belongs to $W_q^{2,1}(Q)$, $q > N+2$, by using the dual estimate for Eu and $E^n u^n$ we can prove that

$$\|(u^n-u-d_n)^+\|_{C^0(0,T;C^{0,1}(\bar{G})) \cap C^{0,1/2}(0,T;C^0(\bar{G}))} \le \frac{M(q)}{3+\omega(N)} \quad (2.22)$$

Now, if $P_n \equiv (x_n,t_n) \notin F_n$; from (2.19) it follows that

$$(u^n-u-d_n)^+(P_n) < \frac{n^{-\delta+\varepsilon}}{\lambda^n+\bar{a}_0} \cdot \frac{M(q)}{3+\omega(N)}. \quad (2.23)$$

If, on the contrary, $P_n \in F_n$ we can prove that there exists a point $\bar{P}_n \equiv (\xi_n,\tau_n) \notin F_n$ "near enough" to P_n, that is :

$$\begin{cases} |x_n-\xi_n| \leqslant \omega(N) \cdot \dfrac{n^{-\delta+\varepsilon}}{\lambda^n+\bar{a}_o} \\ |t_n-\tau_n| \leqslant \left(\dfrac{n^{-\delta+\varepsilon}}{\lambda^n+\bar{a}_0}\right)^2. \end{cases} \quad (2.24)$$

Then, by using (2.22), and (2.24) we have

$$\begin{aligned}(u^n-u-d_n)^+(P_n) &\le (u^n-u-d_n)^+(\bar{P}_n) + \\ &+ \frac{M(q)}{3+\omega(N)}\,(1+\omega(N))\,\frac{n^{-\delta+\varepsilon}}{\lambda^n+\bar{a}_o}\,.\end{aligned} \quad (2.25)$$

Hence, in any case, by (2.23) and (2.25) we have :

$$(u^n-u-d_n)^+ \le \frac{M(q)}{3+\omega(N)}\,(2+\omega(N))\,\frac{n^{-\delta+\varepsilon}}{\lambda^n+\bar{a}_0}\,. \quad (2.26)$$

Analogously we can prove that :

$$(u-u^n-d_n)^+ \le \frac{M(q)}{3+\omega(N)}\,(2+\omega(N))\,\frac{n^{-\delta+\varepsilon}}{\lambda^n+\bar{a}_o}\,, \tag{2.27}$$

from (2.26), (2.27) and (2.15) it follows that

$$\|u^n-u\|_{L_\infty} \le \frac{\lambda^n}{\lambda^n+\bar{a}_o}\,\|u^n-u\|_{L_\infty} + M(q)\,\frac{n^{-\delta+\varepsilon}}{\lambda^n+\bar{a}_o}\,,$$

and the conclusion follows.

We now consider the generalized solution.

The following theorems provides existence, uniqueness and regularity of the generalized solution of (1.5).

<u>Theorem</u> 2.3. *Let us suppose* (1.6) *with* $q \in (\frac{N+2}{2},+\infty)$ *then*

$$\psi_i \in C^o(\bar{Q}) \tag{2.28}$$

$$\begin{cases} \psi_1 < \psi_2 \text{ on } \bar{Q} \\ \psi_1|_\Sigma < 0 < \psi_2|_\Sigma \\ \psi_1(.,0) < 0 < \psi_2(.,0) \text{ on } G. \end{cases} \tag{2.29}$$

Then there exists a unique generalized solution u *of* (1.5) *and we have :*

$$u \in C^o(\bar{Q}),\ u(.,0) = 0 \text{ on } G,\ u|_\Sigma = 0 \tag{2.30}$$

$$\begin{cases} Eu = f \text{ in } D \\ u \in W_q^{2,1}(Q') \end{cases} \tag{2.31}$$

where Q' *is a cylinder* $G' \times]T_1,T_2[$, $0 \le T_1 < T_2 \le T$, G' *is an open regular subset of* G *and* $\bar{Q}' \subset D$,

$$D = \{(x,t) \in \bar{Q},\ \psi_1(x,t) < u(x,t) < \psi_2(x,t)\}.$$

We consider now the Hölder-continuous obstacles and we enounce two regularity results : if the coefficients a_{ij} are merely continuous we can easily prove that the generalized solution of (1.5) is Hölder-continuous with De Giorgi-Nash exponent. If also the coefficients a_{ij} belong to a class of Hölder functions, for instance to the same class of the obstacles, then the generalized solution belongs to same class. More precisely we have :

Theorem 2.4.

1) *Under the hypotheses* (1.2), (2.29),

$$\psi_i \in C^{0,\alpha,\alpha/2}(\bar{Q}),\ \alpha \in (0,1) \text{ and} \tag{2.32}$$

$$f \in L_q(Q),\ q \in (N+1,+\infty)\ ; \tag{2.33}$$

the generalized solution of (1.5) *belongs to*

$$C^{0,\alpha',\alpha'/2}(\bar{Q}), \alpha' = \frac{N}{N+1} \wedge \bar{\alpha} \wedge \alpha,$$

where $\bar{\alpha}$ *is the Hölder exponent of the local solution of the equation relative to the operator* E *(see* [10]*).*

2) *Under the hypotheses* (2.29), (2.32)

$$a_{ij} \in C^{0,\delta,\delta/2}(\bar{Q}),\ \delta \in (0,1), \tag{2.34}$$

and

$$f \in L_q(Q),\ q \in \left(\frac{N+2}{2-\delta \wedge \alpha},+\infty\right)\ ; \tag{2.35}$$

the generalized solution of (1.5) *belongs to* $C^{0,\alpha',\alpha'/2}(\bar{Q})$, $\alpha' = \alpha \wedge \delta$.

See [8] for the proof.

By using theorem 2.3 we obtain the following approximation result for the generalized solution of (1.5).

Theorem 2.5. *Assume* E^n *as in* (2.4) *and* (2.6), (2.32) $f \in L_q$, *with* q *as in* (2.5). *Let* u^n *and* u *be the generalized solution of* $(1.5)_n$ *and* (1.5) *respectively, then*

$$\|u^n-u\|_{L^\infty} \leq cn^{-\frac{\alpha}{2}(\delta-\varepsilon)}, \tag{2.36}$$

where

$$\varepsilon > \frac{(N+3)(\delta+2-2\beta)}{N+q+3}.$$

Remark 2.4. Under the above smoothness hypotheses for the obstacles and free terms, the previous approximation (and Hölder) results still hold for bilateral or unilateral problems whenever there exists a unique generalized solution that can be obtained as a uniform limit of strong solutions. For our problem this is provided by (2.29) or by $(2.11)_i$ and $(2.11)_{ii}$ in [8] ; in the unilateral case there always exists a unique generalized solution, uniform limit of strong ones, see [14].

3. EXTENSION TO THE NON LINEAR CASE.

We are also interested in non variational bilateral problems for quasi-linear parabolic operators.

We consider a Nemytsky operator H associated with a function h(x,t,u,p) with quadratic growth in the gradient variable p. More precisely we assume H(u) = (.,.,u Du), where $h(x,t,r,p) : \bar{Q} \times R \times R^N \to R$ satisfies the following conditions :

$$\begin{cases} \text{i) } h \text{ is continuous with respect to } (x,t,r,p) \in \bar{Q} \times R \times R^N. \\ \text{ii) } |h(x,t,r,p)| \leq k_1 + k_2|p|^2,\ (x,t) \in Q, \\ \qquad |r| \leq C,\ p \in R^N. \\ \text{iii) } h(x,t,r,p) - h(x,t,r',p) \leq k_3(r-r')\ \forall r \geq r' \\ \qquad |r| \leq C,\ |r'| \leq C. \end{cases} \tag{3.1}$$

where k_i are suitable constants possibly depending on C.

The problem we are interested in is : to find u(x,t) s.t. :

$$\begin{cases} u \in W^{2,1}_q(Q), \quad q \in (N+2,+\infty) \\ (Eu-Hu)(u-\psi_1) \le 0, \text{ a.e. in } Q \\ (Eu-Hu)(u-\psi_2) \le 0, \text{ a.e. in } Q \\ u(x,0) = 0 \text{ in } G \\ u|_\Sigma = 0 \end{cases} \tag{3.2}$$

Operators of this kind arise for instance in stochastic problems when a continuous control occurs, see e.g. [2]. In [15] it is proved that under hypotheses (1.1), (1.2), (1.3), (2.1) and (3.1) there exists a unique solution u of problem (3.2).

We consider "linearizing" operators $H^n(u) = h^n(..,u\ Du)$ s.t.

$$h^n(x,t,r,p) \text{ satisfying (3.1).} \tag{3.3}$$

We denote by $(3.2)_n$ the problem (3.2) with $E^n - H^n$ replacing E - H, and by u^n the solution of $(3.2)_n$. As in the linear case we want to evaluate, in the uniform norm, the order of convergence of the solutions of the $(3.2)_n$'s to the solution of (3.2) in terms of the order of convergence of the operators.

Of course we have :

$$\|u\|_{L_\infty(Q)} \vee \|u^n\|_{L_\infty(Q)} \le \|\psi_1\|_{L_\infty(Q)} \vee \|\psi_2\|_{L_\infty(Q)} \equiv M_1. \tag{3.4}$$

By the existence theorem of [15] it follows that u, u^n belong to $L(0,T;W^1(G))$ and

$$\|Du\|_{L_\infty(Q)} \vee \|Du^n\|_{L_\infty(Q)} \equiv M_2. \tag{3.5}$$

Remark 3.1. We can easily prove that it is possible without loss of generality to assume :

$$\sigma \equiv \bar{a}_0 - k_3(\|u\|_{L_\infty(Q)}) > 0 \text{ (see also remark 1.1).} \tag{3.6}$$

Assume that for h^n the following additional hypothesis holds

$$\begin{cases} \text{i) } |h^n(.,.,r,p) - h^n(.,.r',p')| \le k_3|r-r'| + \\ \quad + k_4 n^{1-\eta}|p-p'|, \quad \eta \in [0,1], \end{cases} \tag{3.7}$$

$$\text{ii) } h^n(.,.r,p)-h(.,.r,p) \le k_5 n^{-\gamma}, \ \gamma \in (0,1],$$

with $|r|,|r'| \le C$, $p,p' \in R^N$, $k_4 = k_4(C,|p|)$, $k_5=k_5(C,|p|)$ and k_3 given in (3.1).

We have the following approximation result.

Theorem 3.1. *Assume* (1.1), (1.2), (1.3), (2.4), (2.6), (3.1), (3.3), *and* (3.7) ; *suppose* (2.1) *and* (1.6) *with*

$$q \in ((N+2) \vee \frac{(N+3)(1-\beta\wedge\eta)2}{\delta},+\infty). \tag{3.8}$$

Let u^n *and* u *be the solutions of* $(3.2)_n$ *and* (3.2) *respectively. Then we have :*

$$\|u^n-u\|_{L_\infty(Q)} \le \frac{M(q)}{\sigma} n^{-\delta\wedge\gamma+\varepsilon}, \tag{3.9}$$

where $\sigma = \bar{a}_0 - k_3(\|u\|_{L_\infty(Q)})$ *(see Remark* 3.1),

$$\varepsilon > \frac{(N+3)[\delta\wedge\gamma-2(1-\beta\wedge\eta)]}{N+3+q}$$

and

$$M(q) = c(q)\{\bigvee_{j=1}^{r}\|E\psi_1^j\|_{L_q(Q)} + \bigvee_{j=1}^{s}\|E\psi_2^j\|_{L_q(Q)} + k_6\} \vee k_5$$

with

$$k_6 = k_6(M_1,M_2) \equiv k_1(M_1) + k_2(M_1)M_2^2$$

and M_1,M_2 *given in*(3.4) *and in* (3.5).

Of course in Theorems 2.2, 2.5 and 3.1 it is possible to exchange the roles of ε and q.

Remark 3.2. The problem of defining suitable generalized solutions for problem (3.2), where the strong solution doesn't exist, is open.

For some examples showing how we can "approximate" the therm h(   ), we refer to [9], Theor. 3.1 and 3.2. The numerical approximation for non-variational problems has been studied for example in [4]. There the elliptic linear unilateral case has been considered ; the solution of the non-variational problem is approximated by the affine finite element solution of the regularized variational inequality. Convergence results and error estimates in the L^∞-norm are obstained for such an approximation in the case of Hölder continuous coefficients ($a_{ij} \in C^{0,\delta}(\bar{\Omega})$, $0 < \delta < 1$).

The L^∞-estimate (see (2.7) in Theorem 2.2) can be used to study the convergence of the free-boundary by following some ideas of [3] and [6], as we will show in a forthcoming paper.

REFERENCES.

[1] A. Bensoussan, J.L. Lions. Applications des inéquations variationnelles en contrôle stochastique. Vol. I, Dunod, Paris (1978).

[2] A. Bensoussan, J. Frehse, U. Mosco. A stochastic impulse control problem with quadratic growth Hamiltonian and the corresponding quasi variational inequality. J. Reine Angew. Math. 331 (1982) pp. 124-145.

[3] F. Brezzi, L.A. Caffarelli. Convergence of the Discrete Free Boundaries for Finite Element Approximations. To appear on R.A.I.R.O.

[4] S. Finzi Vita. A numerical approach to the solution of unilateral problems of non-variational type. To appear.

[5] A. Friedman. Stochastic differential equations and applications. Vol. II, Acad. Press, New York (1975).

[6] A. Friedman. Variational Principle and Free-Boundary Problems. Wiley-Interscience, New York (1982).

[7] M.G. Garroni, M.A. Vivaldi. Bilateral inequalities and implicit unilateral systems of the non-variational type. Manuscripta Math. 33 (1980) pp. 177-215.

[8] M.G. Garroni, M.A. Vivaldi. Bilateral evolution problems of non-variational type : existence, uniqueness, Hölder-regularity and approximation of solutions. Manuscripta Math. 48, (1984), pp. 39-69.

[9] M.G. Garroni, M.A. Vivaldi. Approximation results for bilateral non linear problems of non-variational type. Non-Linear Analysis. Vol. 8, N°4, (1984) pp. 301-312.

[10] O.A. Ladyzenskaya, N.N. Ural'ceva. Estimates of Hölder constants for bounded solutions of second order quasilinear parabolic equations of nondivergent form. Lomi Preprints, Leningrad (1981) pp. 1-35.

[11] J.L. Lions, G. Stampacchia. Variational inequalities. Comm. Pure Appl. Math. 20 (1967) pp. 493-519.

[12] O. Nakoulima. Thèse d'Etat. Université de Bordeaux I (1981).

[13] G. Stampacchia. Formes bilinéaires coercitives sur les ensembles convexes. C.R. Acad. Sci. Paris. Série A. 258 (1964) pp. 4416.

[14] G.M. Troianiello. In a class of unilateral evolution problems. Manuscripta Math. 29 (1979) pp. 353-384.

[15] G.M. Troianiello. On solutions to Quasi-linear Parabolic Unilateral Problems. BUMI (6) 1B (1982) pp. 535-552.

Maria Giovanna GARRONI

Dipartimento di Matematica
Università di Roma "La Sapienza"
Piazzale Aldo Moro 2

I.-00185 ROMA

ITALIA

A HARAUX

Two remarks on hyperbolic dissipative problems

0. INTRODUCTION.

In this paper, naturally related to [3] and [9], I describe in a rather informal fashion two remarks which came to my mind a short time after my studying the work of A.V. Babin and M.I. Vishik ([1] and [2]).

Consider a dissipative hyperbolic equation of the form

$$\begin{cases} \partial_t^2 u - \Delta u + f(u) + g(\partial_t u) = h(t,x) & \text{on } \mathbb{R}^+ \times \Omega \\ u = 0 & \text{on } \mathbb{R}^+ \times \Gamma \end{cases} \tag{0.1}$$

where Ω is a bounded, open domain of $\mathbb{R}^N$ $(N \geq 1)$ with a smooth boundary Γ and g is a non-decreasing function such that $g(0) = 0$.

This kind of equations have been studied by many authors in various contexts with one common motivation : to translate the effect of the dissipative term $g(\partial_t u)$ into *global properties* of the dynamical system generated by (0.1).

In this direction the simplest problem is that of energy decay to 0 when f and h are both zero. Then if for example $g \in C^1(\mathbb{R})$ satisfies

$$0 < \alpha \leq g'(v) \leq M < +\infty, \quad \forall v \in \mathbb{R} ,$$

it has been known for a long time that the energy of *any* solution tends to zero exponentially as $t \to +\infty$. However, this property is not associated to strong monotonicity of the generator associated to (0.1) in the energy space $E = H_0^1(\Omega) \times L^2(\Omega)$ endowed with the *usual* inner product : even if $g(v) = \alpha v$ with $\alpha > 0$ this generator is not strictly monotone in this space. In fact a simple *renorming* of E, by adding ã "mixing term" to the inner product, makes the generator strongly monotone and solves the problem in this simple case. Another important question is boundedness of the energy for $t \geq 0$ when

we consider a trajectory u(t,x) of (0.1) and $h \in L^\infty(\mathbb{R}^+, L^2(\Omega))$.
Even when f = 0 this is not an easy question and in a sense this problem, which has been investigated already in 1956 by G. Prodi, is still partially unsolved now. Moreover, most of the results which have been obtained until now on the global behavior of solutions of (0.1) rely in a fundamental way on a method introduced in 1965 by G. Prouse to solve this boundedness question. This very tricky device is explained for example in [9] in which we use the method to obtain a *global dissipativity property* of (0.1).

While studying the results of [3], I have been immediately convinced, for some heuristical reasons which cannot be explained here, that the following two conjectures should be given a positive answer within a few weeks :

Conjecture 1. There must exist a method, simpler than Prouse's method, to prove boundedness of the energy using the "renorming" argument.

Conjecture 2. At least in "subcritical case", the (E_1,E) - attractor for (0,1) constructed in [3] when h(t,x) = h(x) is in fact an E-attractor.

This paper is devided in two sections

- In section 1, we give the proof of conjecture 1.
- In section 2, we outline a proof of conjecture 2.

1. BOUNDEDNESS VIA THE RENORMING TECHNIQUE.

For simplicity, we consider only the "purely dissipative" equation

$$\begin{cases} u \in C(\mathbb{R}^+, H_0^1(\Omega)) \cap C^1(\mathbb{R}^+, L^2(\Omega)) \\ \\ \partial_t^2 u - \Delta u + g(\partial_t u) = h(t,x) \text{ on } \mathbb{R}^+ \times \Omega \end{cases} \tag{1.1}$$

and we assume (this is not the minimal hypothesis)

$$f \in L^\infty(\mathbb{R}^+, L^2(\Omega)) \tag{1.2}$$

We shall also assume (this is not at all necessary) that g is a continuous non-decreasing function defined on $\mathbb{R}$ with $g(0) = 0$.

In order to prove boundedness of trajectories, the following assumption has been done by all authors :

$$\forall v \in \mathbb{R},\ g(v)v \geq \alpha|v|^2 - C \tag{1.3}$$

where $\alpha > 0$.

Another assumption which has been done in all papers on the subject is

$$\begin{cases} \forall v(x) \in H_0^1(\Omega), \quad g(v(x)) \in H^{-1}(\Omega) \text{ and} \\ \\ \|g(v)\|_{H^{-1}(\Omega)} \leq C_1\left(1+\int_\Omega g(v(x))v(x)dx\right) \end{cases} \tag{1.4}$$

Finally it is usual, in order to prove boundedness of trajectories to (0.2), to work at first on a "regularized equation" and to assume that we have

$$u \in W^{1,1}_{loc}(\mathbb{R}^+, H_0^1(\Omega)) \cap W^{2,1}_{loc}(\mathbb{R}^+, H^{-1}(\Omega)) \tag{1.5}$$

Then the energy bound for weak solutions is derived through a standard completion argument. Throughout this section we will use the following notation :

$$V = H_0^1(\Omega), \quad \|v\|^2 = \int_\Omega |\nabla v(x)|^2 dx, \quad \forall\, v \in V$$

$$H = L^2(\Omega), \quad \|u\|^2 = \int_\Omega |u(x)|^2 dx, \quad \forall\, u \in H$$

$$V' = H^{-1}(\Omega)$$

Also the inner product in H is denoted by (,) and the duality pairing in $V' \times V$ by <,>. Finally the first eigenvalue of $(-\Delta)$ in $H_0^1(\Omega)$ will be denoted by λ_1, and the norm of $h(t,x)$ in $L\,(\mathbb{R}^+, L^2(\Omega))$ will be denoted simply by $\|h\|_\infty$.

We now state our result.

Theorem 1.1. *Assume that* (1.2), (1.3) *and* (1.4) *are satisfied and for any weak solution* u *of* (1.1) *consider the energy*

$$E(t) = \frac{1}{2}\|u(t)\|^2 + \frac{1}{2}|\partial_t u(t)|^2 \quad \textit{for } t \geq 0.$$

Then we have $E(t) \in L^\infty(\mathbb{R}^+)$. *More precisely, there exists two constants* K, L *depending only on* α, C, C_1 *and* Ω *such that if* $\varepsilon > 0$ *is chosen so small that* $\varepsilon \leq \inf\{\frac{\alpha}{6}, \frac{1}{2}\sqrt{\alpha_1}\}$ *and*

$$E(0) \leq \frac{K}{\varepsilon^2} - \frac{L}{\varepsilon}(1+\|h\|_\infty^2) \tag{1.6}$$

then we have

$$\sup_{t\geq 0} E(t) \leq \frac{4K}{\varepsilon^2} \tag{1.7}$$

and

$$\limsup_{t\to+\infty} E(t) \leq \frac{4L}{\varepsilon}(1+\|h\|_\infty^2) \tag{1.8}$$

Proof. For any $\varepsilon > 0$, we set

$$E_\varepsilon(t) = E(t) + \varepsilon\langle u'(t), u(t)\rangle \tag{1.9}$$

Clearly for any ε such that $0 < \varepsilon \leq \frac{1}{2}\sqrt{\lambda_1}$, we have $\varepsilon|\langle u'(t),u(t)\rangle| \leq \frac{\varepsilon}{\sqrt{\lambda_1}}\|u(t)\|\,|u'(t)| \leq \frac{1}{2}E(t)$, hence

$$\frac{1}{2}E(t) \leq E_\varepsilon(t) \leq \frac{3}{2}E(t) \leq 2\,E(t) \tag{1.10}$$

Let u satisfy (1.1) and (1.5). Then

$$E_\varepsilon'(t) = \langle h(t) - g(u'), u'\rangle + \varepsilon|u'|^2 + \varepsilon\langle u'', u\rangle$$

$$= \langle h(t) - g(u'), u'\rangle + \varepsilon|u'|^2 - \varepsilon\|u\|^2 \qquad (1.11)$$

$$- \varepsilon\langle g(u'), u\rangle + \varepsilon\langle h(t), u(t)\rangle$$

For all $t \geq 0$ we set

$$w(t) = \langle g(u'(t)), u'(t)\rangle = \int_\Omega g(u'(t,x))u'(t,x)dx.$$

Then $w \geq 0$ and from (1.3), (1.4) and (1.11) we deduce that a.e. on $\mathbb{R}^+$

$$E_\varepsilon'(t) \leq - \frac{1}{2}w(t) - \frac{\alpha}{2}|u'(t)|^2 + \frac{C}{2} + \varepsilon|u'(t)|^2 - \varepsilon\|u(t)\|^2$$

$$+ C_1\varepsilon w(t)\|u(t)\| + C_1\varepsilon\|u(t)\| + |h(t)||u'(t)| + \varepsilon|h(t)||u(t)|.$$

$$\leq - (\frac{\alpha}{2} - \frac{\alpha}{4} - \varepsilon)|u'(t)|^2 - \frac{\varepsilon}{2}\|u(t)\|^2$$

$$+ (C_1\varepsilon\|u(t)\| - \frac{1}{2})w(t)$$

$$+ M(C_1,\lambda_1,\alpha)(1 + \|h\|_\infty^2)$$

Thus if we assume $\varepsilon \leq \frac{\alpha}{6}$, we obtain

$$E_\varepsilon' \leq - \varepsilon E(t) + w(t)(C_1\varepsilon\|u(t)\| - \frac{1}{2}) + M(1 + \|h\|_\infty^2).$$

Clearly $\|u(t)\| \leq \sqrt{2\ E(t)} \leq 2\sqrt{E_\varepsilon(t)}$ and we finally deduce from (1.10) and the inequality above that

$$E_\varepsilon' \leq - \frac{\varepsilon}{2} E_\varepsilon + w(t)(2\varepsilon C_1\sqrt{E_\varepsilon(t)} - \frac{1}{2}) + H \qquad (1.12)$$

with $H = M(C_1,\lambda_1,\alpha)(1 + \|h\|^2)$

We claim that (1.12) implies our result with

$$K = \frac{1}{2}\left(\frac{1}{4C_1}\right)^2, \quad L = M(C_1,\lambda_1,\alpha) > 0. \qquad (1.13)$$

Indeed, assume that (1.6) is satisfied with K and L given by (1.13) and $\varepsilon \leq \inf\{\frac{\alpha}{6},\frac{1}{2}\sqrt{\lambda_1}\}$.

Then we have :

$$E_\varepsilon(0) \leq 2(\frac{K}{\varepsilon^2} - \frac{H}{\varepsilon}) \tag{1.14}$$

Let us show that (1.14) and (1.12) imply

$$E_\varepsilon(t) \leq \frac{2K}{\varepsilon^2}, \quad \forall t \geq 0 \tag{1.15}$$

Indeed, if (1.15) is not satisfied, let

$$t_o = \mathrm{Inf}\{t \geq 0, \quad E_\varepsilon(t) > \frac{2K}{\varepsilon^2}\}$$

Clearly we have $t_o > 0$ and also

$$E_\varepsilon(t) \leq \frac{2K}{\varepsilon^2}, \quad \forall t \in [0,t_o] \tag{1.16}$$

$$E_\varepsilon(t_o) = \frac{2K}{\varepsilon^2} \tag{1.17}$$

From (1.12), (1.13) and (1.16) we deduce

$$E'_\varepsilon \leq -\frac{\varepsilon}{2} E_\varepsilon + H, \text{ a.e. on } [0,t_o] \tag{1.18}$$

And it is immediate to deduce from (1.18) that

$$E_\varepsilon(t) \leq \exp(-\frac{\varepsilon}{2}t)\, E_\varepsilon(0) + \frac{2H}{\varepsilon}, \quad \forall t \in [0,t_o] \tag{1.19}$$

From (1.14) and (1.19) applied with $t = t_o$ we derive $E_\varepsilon(t_o) < 2(\frac{K}{\varepsilon^2} - \frac{H}{\varepsilon}) + \frac{2H}{\varepsilon} = \frac{2K}{\varepsilon^2}$.

This contradicts (1.17). Hence $t_o = +\infty$ and we have (1.15). From (1.15) and (1.10) inequality (1.7) follows immediately.

Finally (1.19) actually holds true with $t_0 = +\infty$ and by letting $t \to +\infty$ in (1.19) we get :

$$\limsup_{t\to+\infty} E_\varepsilon(t) \le \frac{2H}{\varepsilon} .$$

This inequality together (1.10) gives (1.8).

Hence theorem 1.1 is completely proved.

Remark 1.2. a) Clearly (1.6) is satisfied for any $\varepsilon > 0$ *small enough*. Hence we actually proved that $E(t) \in L^\infty(\mathbb{R}^+)$.

b) From theorem 1.1 it would be easy to deduce that for some constant $R \ge 0$ depending only on α, C, C_1 and Ω, we have

$$\limsup_{t\to+\infty} E(t) \le R(1 + \|h\|_\infty^4) \tag{1.20}$$

for any solution u of (1.1).

c) If $g(v) = c_1 v + c_2 |v|^{\gamma-1} v$ with $c_1 \ge 0$ and $c_2 > 0$, condition (1.4) is automatically satisfied if either $N \le 2$ or $N \ge 3$ and $\gamma \le \frac{N+2}{N-2}$.

In such a case (1.20) is *not optimal* and a careful analysis then shows that in fact

$$\limsup_{t\to+\infty} E(t) \le N'(1 + \|h\|_\infty^2)$$

We do not know whether such a refinement of (1.20) is available for a general monotone function g satisfying (1.3) and (1.4).

d) It is still unknown whether trajectories of (1.1) satisfy $E(t) \in L^\infty(\mathbb{R}^+)$ when $g(v) = |v|^{\gamma-1} v$ with $N \ge 3$ and $\gamma > \frac{N+2}{N-2}$. Our method fails at exactly the same level where all previous attempts in this direction have failed since 1956.

2. ON A RESULT OF A.V. BABIN AND M.I. VISHIK.

We now consider the nonlinear autonomous equation

$$\begin{cases} u \in C(\mathbb{R}^+, H_0^1(\Omega)) \cap C^1(\mathbb{R}^+, L^2(\Omega)) \\ \partial_t^2 u - \Delta u + \varepsilon \partial_t u + f(u) = h(x) \text{ on } \mathbb{R}^+ \times \Omega \end{cases} \tag{2.1}$$

where $\varepsilon \in 0$, $h \in L^2(\Omega)$ and $f \in C^1(\mathbb{R})$.

Let $E = H_0^1(\Omega) \times L^2(\Omega)$ and assume $N \leq 3$. Then (2.1) generates a nonlinear semi-group of continuous transformations $S_t : E \to E$ as soon as f satisfies the following conditions

$$\begin{aligned} |f'(u)| &\leq C(1+|u|^k) \text{ for some } k \geq 0 \text{ if } n=2 \\ |f'(u)| &\leq C(1+u^2) \qquad\qquad \text{if } n=3 \end{aligned} \tag{2.2}$$

in addition to some "sign" conditions, for example

$$\operatorname*{Inf}_{u \in \mathbb{R}}\{f'(u)\} > -\infty, \quad \operatorname*{Inf}_{u \in \mathbb{R}}\{uf(u)\} > -\infty \tag{2.3}$$

Moreover, under the conditions (2.2) and (2.3), it is known (cf. for example [7], Theorem 2.2) that any trajectory $(u(t), \partial_t u(t)$ of (2.1) must lie entirely in a *fixed* bounded set of E for $t \geq t_0$ where t_0 depends on the initial data $(u(0), \partial_t u(0))$. In fact, either a direct use of the nonorming technique or a careful study of the estimates in [7] shows that t_0 only depends of the energy (or the norm in E) of the initial state $(u(0), \partial_t u(0))$. Hence we have

Proposition 2.1. Under the above assumptions, there exists a bounded set $B_0 \subset E$ such that for any bounded $B \subset E$, there exists $t(B)$ with the following property :

$$\forall t \geq t(B),\ S_t(B) \subset B_0 \tag{2.4}$$

This result suggests that the nonlinear semi-group $\{S_t\}$ may have an "attractor" in the sense of E. Unfortunately, it is well-known that even in the case f = 0, h = 0, the operator $S_t : E \to E$ is *not compact* for any $t > 0$. Hence the general results on the existence of a maximal attractor cannot be used. In fact, there *cannot* exist any compact "absorbing" set for S_t in E.

This difficulty has been circumvented by A.V. Babin - M.I. Vishik who introduced the idea of an (E_1,E)-attractor (cf. [1] and [3]). More precisely, let $E_1 = H^2 \cap H_0^1(\Omega) \times H_0^1(\Omega)$. In [3], the following result is established.

Theorem 2.2. *Under the above assumptions, there exists* $\mathcal{A}$ *a closed and bounded subset of* E_1 *such that*

$$\forall t \geq 0, \quad S_t\mathcal{A} = \mathcal{A} \tag{2.5}$$

$\forall B_1$ *bounded in* E_1, *we have*

$$\operatorname*{Sup}_{w \in B_1} \operatorname{dist}_E(S_t w, \mathcal{A}) \underset{t \to +\infty}{\to} 0 \tag{2.6}$$

It is clear that such a set $\mathcal{A}$ must be unique. In [3], it is proved that $\mathcal{A}$ is equal to the unstable set generated by the set $\mathcal{M}$ of stationary solutions cf (2.1) in E (or E_1, these two sets being equal).

The proof of this result relies on non-standard estimates, especially on the subtle remark of [1] that $S_t : E_1 \to E_1$ is "uniformly bounded". Of course in identifying the attrator $\mathcal{A}$, a crucial role is played by the existence of a Liapunov functional

$$\Phi(u,p) = \int_\Omega \{\tfrac{1}{2}|\nabla u|^2 + \tfrac{1}{2}|p|^2 + F(u) - hu\}dx.$$

The continuity of this functional with respect to weak sequential convergence in E along negative trajectories bounded in E_1 was the original motivation in looking for an (E_1,E)-attractor and getting this elegant result of uniform boundedness for $S_t : E_1 \to E_1$. For me it still remained unnatural to restrict the convergence property (2.6) to subsets B_1 bounded in E_1, and the argument that S_t is not compact was balanced by the idea of some "asymptotic smoothing effect" which does exist when one considers a single trajectory on $\mathbb{R}^+$. As a matter of fact, we shall now establish the following refinement of theorem 2.2.

Theorem 2.3. *Assume that the assumptions in theorem 2.2 are satisfied, with either* $N \leq 2$, *or* $N = 3$ *and*

$$\exists \delta > 0, \ \forall u \in \mathbb{R}, \ |f'(u)| \leq C(1+|u|^{2-\delta}) \tag{2.7}$$

Then in fact we have

$$\forall B \text{ bounded in } E, \ \sup_{w \in B} \operatorname{dist}_E(S_t w, \mathcal{A}) \underset{t \to +\infty}{\to} 0 \tag{2.8}$$

with $\mathcal{A}$ *the same set as in the statement of theorem 2.2 ; and more precisely with the notation of* [3] : $\mathcal{A} = M_+(\mathcal{M}(E)) = M_+^1(\mathcal{M}(E_1))$.

Proof of Theorem 2.3. The argument is rather simple if N=1 and becomes technically delicate when N = 2 or 3. In all cases, it relies on the following general lemma which extends theorem 1.4 of [3].

Lemma 2.4. *Let* E *be any real Banach space and* $S_t : E \to E$ *a semi-group of continuous operators. We assume that there exists a* compact *subset* K *of* E *such that*

$$\forall B \text{ bounded in } E, \ \sup_{u \in B} d(S_t u, K) \underset{t \to +\infty}{\to} 0 \tag{2.9}$$

Them : S_t *has a maximal attractor* $\mathcal{A}$ *in the sense of* E *and* $\mathcal{A} \subset K$.

Proof. For any B bounded in E, we set $\forall \tau > 0$, $B_\tau = \bigcup_{t \geq \tau} S_t(B)$ and $\omega(B) = \bigcap_{\tau > 0} \bar{B}_\tau^{\,E}$. Then we have

(1°) For all B bounded, $\omega(B) \neq \emptyset$ (if $B \neq \emptyset$) and $\omega(B) \subset K$.

Actually it is immediate that $y \in \omega(B)$ if and only if there exists a sequence $t_n \to +\infty$ and points $u_n \in B$ such that $y = \lim_{n\to+\infty} S_{t_n}(u_n)$.

As a consequence of (2.9) we thus have $\omega(B) \subset K$. To prove that $B \neq \emptyset \Rightarrow \omega(B) \neq \emptyset$ it is sufficient to remark that if $u_n \in B$ for all $n \in \mathbb{N}$ and $\{t_n\}$ is any sequence of $\mathbb{R}^+$ with $t_n \to +\infty$, there exists a sequence $a_n \in K$ such that $\lim_{n\to+\infty} \|S_{t_n} u_n - a_n\| \to 0$, as follows from (2.9). Now if $\{n_k\}$ is a sequence of integers tending to infinity such that $a_{n_k} \to \xi$ in K, it is clear that $\xi \in \omega(B)$.

(2°) $\forall\theta > 0,\ S_\theta\omega(B) = \omega(B)$.

The proof parallels that of the same step in the proof of theorem 1.4 of [3].

(3°) If $\mathcal{A} = \omega(K)$, then $\mathcal{A}$ is a (compact) maximal attractor for S_t in E.

Indeed for any B bounded in E, $\omega(B) \subset K$, hence $\omega(B) = S_\theta\omega(B) \subset S_\theta K,\ \forall\theta > 0 \Rightarrow \omega(B) \subset \bigcap_{\theta>0} S_\theta K \subset \omega(K) = \mathcal{A}$. Finally assume that for some B bounded, $\neq \emptyset$ we have $\limsup_{t\to+\infty} \operatorname{dist}(S_t B, \mathcal{A}) = 2\varepsilon > 0$.

Then there exists $t_n \to +\infty$ and $u_n \in B$ such that

$$d(S_{t_n} u_n, \mathcal{A}) \geq \varepsilon \tag{2.10}$$

Now as a consequence of (2.9) we may assume that $S_{t_n} u_n \to \zeta \in K$ as $n \to +\infty$. Then by definition $\zeta \in \omega(B) \Rightarrow \zeta \in \mathcal{A}$. By letting $n \to +\infty$ in (2.10) we have on the other hand : $d(\zeta, \mathcal{A}) \geq \varepsilon > 0$. This contradiction shows that in fact, for any B bounded in E, we must have $\operatorname{Sup}_{u\in B} \operatorname{dist}(S_t u, \mathcal{A}) \to 0$ as $t \to +\infty$.

Since obviously $\mathcal{A}$ is invariant and contained in K, the proof of lemma 2.4 is achieved.

In order to prove theorem 2.3 we need some more preparation. First we introduce the space

$$E' = L^2(\Omega) \times H^{-1}(\Omega)$$

and the linear operator in E' defined by

$$\begin{cases} D(A) = H_0^1(\Omega) \times L^2(\Omega) \\ \\ A(u,p) = (-p, -\Delta u + \varepsilon p), \ \forall (u,p) \in D(A) \end{cases} \tag{2.11}$$

It is straightforward, as a consequence of Lax-Milgram theorem, to check that A is onto. Since A is monotone, it is therefore maximal monotone in E'.

Since A is a bounded perturbation of a skew-adjoint operator in E', the contraction semi-group generated by -A in E' is in fact a *group* which we shall denote by $T(t) : E' \to E'$.

The proofs of the following facts are easy and will therefore be omitted

- The semi-group T(t) is *exponentially damped* in E', which means that there exists $M \in \mathbb{R}^+$ and $\lambda > 0$ such that

$$\|T(t)\|_{\mathcal{L}(E',E')} \leq Me^{-\lambda t} \quad , \quad \forall t \geq 0$$

- We have $T(t)E \subset E$, $T(t)E_1 \subset E_1$, the restriction of T(t) to either E or E_1 is a contraction semi-group for the usual Hilbert structure, and this semi-group is exponentially damped in both spaces.

As a consequence of the first property, for any $g \in L^\infty(\mathbb{R}, H^{-1}(\Omega))$, the evolution equation in $E' = L^2(\Omega) \times H^{-1}(\Omega)$

$$\frac{dU}{dt} + AU(t) = (0, g(t)) \tag{2.12}$$

has one and only one solution $U \in L^\infty(\mathbb{R}, E')$ which is represented by the formula (cf.[6], p.155)

$$\underline{U}(t) = [\mathcal{T}(g)](t) = \int_0^{+\infty} T(\tau)(0, g(t-\tau))d\tau \tag{2.13}$$

The operator $\mathcal{T}$ will be used several times in the proof of theorem 2.3. The following property of $\mathcal{T}$ will be useful in all cases.

Lemma 2.5. *If we assume that* g *satisfies*

$$g \in L^{\infty}(\mathbb{R}, H_0^1(\Omega)) \tag{2.14}$$

then :

$$\mathcal{C}(g) \in C_B(\mathbb{R}, E_1). \tag{2.15}$$

Proof of Theorem 2.3 continued.

a) Case N=1. This case is easier since we can establish in *one step* that (2.9) is satisfied with K a *bounded subset of* E_1.

Indeed, if B is bounded in E, according to proposition 2.1, there exists t(B) such that $\forall t \geq t(B)$, $S_t(B) \subset B_0$. Hence for $t \geq t(B)$, any solution of (2.1) with initial data $(u(0),(\partial_t u)(0))$ in B is such that $u(t) \in pr_1(B_0)$, a bounded subset of $H_0^1(\Omega)$.

Now we introduce $g(t) \in L^{\infty}(\mathbb{R}, H_0^1(\Omega))$ defined by

$$g(t,x) = \begin{cases} -f(u(t(B),x)) & \text{for } t \leq t(B) \\ -f(u(t,x)) & \text{for } t \geq t(B). \end{cases}$$

As a consequence of lemma 2.5, we have $\mathcal{C}(g) \in C_B(\mathbb{R}, E_1)$ and moreover for all initial data in B, the function $\mathcal{C}(g)$ remains in a *fixed* bounded subset B_1 of E_1 for all $t \in \mathbb{R}$.

Let ψ be the solution of the elliptic problem

$$\psi \in H^2(\Omega) \cap H_0^1(\Omega), \ -\Delta\psi = h(x).$$

Clearly we have $(\psi(x),0) = \mathcal{C}(h(x))$.

Now for all $t \geq t(B)$, we have

$$U(t) = T(t-t(B))\{U(t(B)) - \underline{U}(t(B))\} + \underline{U}(t) \tag{2.16}$$

with $\underline{U}(t) = \mathcal{T}(g(t,x)+h(x))(t)$ which remains in a *fixed* bounded subset B_2 of E_1.

Let $K = \bar{B}_2^E$. Then K is compact in E and bounded in E_1.

From (2.16) and the fact that T(t) is exponentially damped in E, we deduce

$$d(U(t),K) \le \|U(t)-\underline{U}(t)\|_E \le C\exp[-\nu(t-t(B))] \le C_1\exp(-\nu t),$$

As a consequence of lemma 2.4 :
S_t has a maximal attractor $\mathcal{A}_E$ in the sense of E, and since $\mathcal{A}_E \subset K$, it is a bounded subset of E_1. Now if we apply (2.9) with $K = \mathcal{A}_E$ and B a bounded subset of E_1; we obtain that $\mathcal{A}_E$ is an (E_1,E) attractor. By uniqueness of such an attractor it follows that $\mathcal{A}_E = \mathcal{A}$.

b) Case $N > 1$. We start as previously but since the operator $v \to f \circ v$ does not carry $H_0^1(\Omega)$ into itself we do not have $g(t) \in H_0^1(\Omega)$ with g(t) as above.

We can, however, establish that $\underline{U}(t)$ remains in a compact subset of E, independant of B, through the following abstract device.

First of all we notice that the operator $u \to f \circ u$ is continuous from $L^2(\Omega)$ into itself when restricted to the compact set $pr_1(B_0)$, hence it is uniformly continuous. Moreover for $t \ge t(B)$, any solution $U(t) = (u(t),p(t))$ starting in B at $t = 0$ is such that $|u_t|_{L^2(\Omega)}$ remains a priori bounded.

It follows that the functions $t \to g(t,x)$ defined above are confined to a uniformly equicontinuous family of functions : $\mathbb{R} \to L^2(\Omega)$.

Now the linear operator $\mathcal{T}: g \to \underline{U}$ is bounded from $L^\infty(\mathbb{R},L^2(\Omega))$ to $C_B(\mathbb{R},E)$. It follows obviously from the property just above that the trajectories $\underline{U} : \mathbb{R} \to E$ are uniformly equicontinuous (With a continuity modulus which does not depend on B).

By using the equation satisfied by $\underline{U}(t) = (\underline{u}(t),\underline{p}(t))$, the same argument as in [8], lemma 4.3 and p.210 finally gives that $\bigcup_{t\in\mathbb{R}} \{\underline{U}(t)\}$ is confined to some precompact subset C in E. On using (2.16) as in case N=1, we obtain that (2.9) is satisfied with K some compact subset of E.

Hence, by lemma 2.4, there exists $\tilde{\mathcal{A}}$ compact in E and such that

$$\forall t \geq 0, \quad S_t \tilde{\mathcal{A}} = \tilde{\mathcal{A}} \tag{2.17}$$

$$\forall B \text{ bounded in } E, \ \sup_{u \in B} d(S_t u, \tilde{\mathcal{A}}) \to 0 \text{ as } t \to +\infty \tag{2.18}$$

Now the proof of theorem 2.3 will be completed as in case N=1 if we prove the following for $N \geq 2$.

Theorem 2.6. *Let* $\tilde{\mathcal{A}}$ *be any bounded subset of* E *which satisfies* (2.17). *Then in fact* $\tilde{\mathcal{A}}$ *is a bounded subset of* E_1.

Sketch of proof of theorem 2.6. The conclusion of theorem 2.6 is equivalent to obtaining an *a priori* bound in $L^\infty(\mathbb{R},E_1)$ of complete trajectoires of S_t with range contained in $\tilde{\mathcal{A}}$. By setting $p = \partial_t u$ and differentiating the equation for u we obtain

$$\partial_t^2 p - \Delta p + \varepsilon p = -f'(u)p = g(t) \tag{2.19}$$

In fact since $p \in L^\infty(\mathbb{R},L^2(\Omega)) \cap W^{1,\infty}(\mathbb{R},H^{-1}(\Omega))$, the vector $V(t) = (p(t),\partial_t p(t))$ satisfies the equation

$$V(t) = \mathcal{T}(-f'(u)p) = \mathcal{T}(g) \tag{2.20}$$

whereas $U(t) = (u(t),\partial_t u(t))$ is the solution of

$$U(t) = \mathcal{T}(-f(u(t,x)+h(x)) \tag{2.21}$$

Clearly when N=1, theorem 2.6 is a consequence of theorem 2.3 which is already proved in this case.

Now we prove theorem 2.6 for $N \in \{2,3\}$.

For this purpose we introduce some notation and a lemma which is an easy consequence of interpolation theory (cf. [10]).

Notation : we denote by Λ the self-adjoint unbounded operator in the Hilbert space $L^2(\Omega)$ defined by

$$D(\Lambda) = H^2(\Omega) \cap H_0^1(\Omega) \ ; \quad \Lambda u = -\Delta u, \quad \forall u \in D(\Lambda).$$

For all $\sigma \geq 0$ the domain of the "fractional power" Λ^σ will be denoted by D_σ.

For any $\phi \in H^{-1}(\Omega)$, the solution ψ of $\psi \in H_0^1(\Omega)$, $-\Delta\psi = \phi$ will be represented by $K\phi$ with $K \in \mathcal{L}(H^{-1}(\Omega), H_0^1(\Omega))$.

Lemma 2.7. *The operators* Λ *and* K *have the following properties*

a) $\forall \rho \in [0,1]$ *such that* $\rho \neq \frac{1}{2}$, *we have*

$$K(H^{-1+\rho}) \subset D_{\frac{1}{2}+\frac{\rho}{2}} = D(\Lambda^{\frac{1+\rho}{2}})$$

b) $\forall \sigma \geq 0, \quad D_\sigma \subset H^{2\sigma}(\Omega)$

c) $\forall \sigma \geq 0,$ *if* $g \in L^\infty(\mathbb{R}, D_\sigma)$ *then* $\mathcal{T}(g(t)) \in L^\infty(\mathbb{R}, D_{1+\sigma} \times D_\sigma)$

a) Case N=2. Let $\delta \in]0,1[$: since U(t) is *a priori* bounded in $L^\infty(\mathbb{R},E)$, g(t) is *a priori* bounded in $L^\infty(\mathbb{R},L^{2-\delta})$. For δ small enough we deduce that g(t) is bounded in $L^\infty(\mathbb{R},H^{-\rho}(\Omega))$ with $\rho \in]0,\frac{1}{2}[$.

Then K g(t) is bounded in $D(\Lambda^{1-\frac{\rho}{2}})$ and therefore $(Kp(t), K\partial_t p(t)) = \mathcal{T}(Kg(t))$ is bounded in $L^\infty(\mathbb{R},D_{1+\sigma} \times D_\sigma)$ with $\sigma = 1 - \frac{\rho}{2} > \frac{3}{4}$. This implies that we have

$$K\partial_t p(t) \in L^\infty(\mathbb{R},C(\bar{\Omega}))$$

On the other hand, since K commutes with $(-\Delta)$ we find

$$u(t,x) = \psi(x) - K(f(u(t,x)) - \varepsilon p(t,x) - \partial_t p(t,x)) \qquad (2.22)$$

hence u is *a priori* bounded in $L^\infty(\mathbb{R},C(\bar{\Omega}))$, then f(u) is a priori bounded in $L^\infty(\mathbb{R},H_0^1(\Omega))$ and by using (2.21) we conclude that U(t) is a priori bounded in

$C_B(\mathbb{R},E_1)$ as a consequence of lemma 2.5.

b) Case N=3. The only thing which differs from the case N=2 above is how to get an *a priori* bound of u(t,x) in $L^\infty(\mathbb{R},C(\bar{\Omega}))$. This bound is more and more difficult to obtain as δ in inequality (2.7) becomes smaller.

The idea is to perform a "bootstrap" argument using (2.20), (2.22) and lemma 2.7, in addition to the fact that for any $q \in [1,2]$, we have

$$L^q(\Omega) \subsetneq H^{-s}(\Omega), \; s = 3(\tfrac{1}{q}-\tfrac{1}{2}) \tag{2.23}$$

Starting from an estimate of u in $L^k(\Omega)$ with $k \geq 2^* = 6$ and setting $2 - \delta = r$ in (2.7), we get f'(u)p estimated in $L^q(\Omega)$ with $\frac{1}{q} = \frac{r}{k} + \frac{1}{2}$.

This shows that g(t) is bounded in $H^{-s}(\Omega)$ with $s = \frac{3r}{k}$. As long as s is kept $\neq \frac{1}{2}$, through lemma 2.7 we deduce that u is bounded in $H^{2-s}(\Omega)$.

- If $r < 1$, in fact we find $s < \frac{1}{2}$ by choosing $k = 6$: in this case $H^{2-s}(\Omega) \subsetneq C(\bar{\Omega})$ and the L^∞-bound is obtained in *one step*.

- If $r = 1$, we obtain in one step that u is bounded in $H^{\frac{3}{2}-\alpha}(\Omega)$ for any $\alpha > 0$. In this case a *second step* like the proof in case N = 2 gives the bound in $L^\infty(\Omega)$ (observe that from the first step we deduce that g(t) is bounded in $L^\infty(\mathbb{R},L^{2-\delta}(\Omega))$).

- If $r > 1$, we define a sequence $\{k_n\}$ inductively by the formula

$$\frac{1}{k_{n+1}} = \frac{1}{2} - \frac{1}{3}(2-\frac{3r}{k_n}) = -\frac{1}{6} + \frac{r}{k_n} \tag{2.24}$$

and $k_1 = 6$.

This sequence will be stopped at the last index n_0 such that $k_n \leq 6r$ (Note that k_n is increasing and n_0 is always finite).

If $k_{n_0} < 6r$, then k_{n_0+1} defined by (2.24) is finite, positive and $k_{n_0+1} > 6r$. Thus after n_0+1 steps, thanks to lemma 2.7 we get u estimated in

$H^{\frac{3}{2}+\eta}(\Omega)$ with $\eta > 0$.

In the exceptional situation where $k_{n_0} = 6r$, we conclude like in case $r = 1$ above.

End of proof of theorem 2.3. Let $\tilde{\mathcal{A}}$ satisfy (2.17) and (2.18) with $\tilde{\mathcal{A}}$ compact in E. As a consequence of theorem 2.6, $\tilde{\mathcal{A}}$ is a bounded subset of E_1. This together with (2.17) and (2.18) implies that $\tilde{\mathcal{A}}$ is an (E_1,E)-attractor for S_t, hence $\tilde{\mathcal{A}} = \mathcal{A} = M_+^1(\mathcal{M}(E))$. On the other hand theorem 2.6 implies that $M_+(\mathcal{M}(E))$ is equal to $M_+^1(\mathcal{M}(E))$, since the closure of the range of a complete trajectory which is bounded as $t \to -\infty$ is a bounded subset of E which satisfies (2.17).

Remark 2.8.

- We do not know what happens when N=3 and $\delta = 0$ in (2.7). At least in this case there is an (E_1,E)-attractor (cf. [3]).

- Theorem 2.3 shows that, like in the case of parabolic dissipative systems, $\{S_t\}$ has a *global* asymptotic "smoothing effect" in E as $t \to +\infty$. But the situation is more delicate than for parabolic systems since here there is no smoothing effect of global type for finite values of t (i.e. S_t is *not* compact : $E \to E$ even if $f = h = 0$).

REFERENCES.

[1] A.V. Babin, M.I. Vishik, Regular attractors of semi-groups and evolution equations, J. Math. pures et appl. 62 (1983), 441-491.

[2] A.V. Babin, M.I. Vishik, Attractors of some evolution equations and estimates of their dimension, Uspekhi mat. nauk 38 (1983), 133-187 (in russian)

[3] A.V. Babin, M.I. Vishik, Attracteurs maximaux dans les équations aux dérivées partielles, this volume.

[4] J.M. Ball, On the asymptotic behavior of generalized processes, with applications to nonlinear evolution equations, J. Diff. Eq. 27 (1978), 224-265.

[5] C.M. Dafermos, Almost periodic processes and almost periodic solutions of evolution equations, in Dynamical Systems, Proc. Univ. Florida International Symposium, Academic Press (1977), 43-57.

[6] A. Haraux, Nonlinear Evolution Equations - Global behavior of solutions, Springer-Verlag, Lecture Notes in Math n°841 (1981).

[7] A. Haraux, Dissipativity in the sense of Levinson for a class of second-order nonlinear evolution equations, Nonlinear Analysis, T.M.A. 6 (11) (1982), 1207-1220.

[8] A. Haraux, Almost periodic forcing for a wave equation with a non-linear, local damping term, Proc. Roy. Soc. Edinburgh 94A (1983), 195-212.

[9] A. Haraux, Some new results on nonlinear wave equations in a bounded domain, in Nonlinear partial differential equations and their applications, Collège de France Seminar, volume IV, Pitman, Research Notes in Math n°84 (1982), 137-147.

[10] J.L. Lions, E. Magenes, Problèmes aux limites non homogènes et applications, volume 1, Dunod, Paris (1968).

[11] G. Prodi, Soluzioni periodiche di equazioni a derivati parzioli di tipo iperbolico non lineari, Ann. Mat. Pura ed Appl. 42 (1956), 25-49.

[12] G. Prouse, Soluzioni quasi periodiche della equazione delle onde con termine dissipathio non lineare, I, II, III, IV, Atti. Accad. Naz. Lineari Rc. 38-39 (1965).

Alain HARAUX

Université Pierre et Marie Curie - CNRS
Laboratoire Analyse Numérique
Tour 55-65 - 5e étage
4 place Jussieu

75230 PARIS CEDEX 05

FRANCE

V A KONDRATIEV & O A OLEINIK

On the smoothness of weak solutions of the Dirichlet problem for the biharmonic equation in domains with nonregular boundary

1. Smoothness of weak solutions of the Dirichlet problem for the biharmonic equation in two independent variables has been considered by many authors. In papers [I]-[3] unimprovable estimates of the continuity modulus at a boundary point are obtained for a weak solution and its first derivatives with regard for geometric properties of the boundary in a neighbourhood of this point. Under certain conditions it is also proved that a weak solution in a closed domain belongs to the Hölder class $C^{1+\delta(\omega)}$, where $\delta(\omega)$ is a root of a transcendental equation with a parameter determined by the geometric properties of the domain ; and it is shown that the Hölder class $C^{1+\delta(\omega)}$ can not be improved for domains characterized by the same values of the parameter ω. As a consequence of the theorems for the biharmonic equation, similar results are obtained in [3] for the von Karman equations and the plan Navier-Stokes system.

In this paper we consider the biharmonic equation with the Dirichlet homogeneous boundary conditions and with distributions in the right-hand side belonging to a wider class than in [2]-[3]. This allows to study the regularity and the behaviour of weak solutions of the Dirichlet nonhomogeneous boundary value problem. The restriction imposed on the right-hand side of the biharmonic equation are in some sense necessary for the obtained results to be valid.

We introduce the following notation. Let Ω be a bounded domain in (x_1,x_2) plane. By $\overset{\circ}{H}{}^2(\Omega)$ we denote the completion of $C_0^\infty(\Omega)$ (the space of infinitely differentiable functions with a compact support in Ω) in the norm

$$\|u\|_2 = \Big(\sum_{|\alpha|\leq 2} \int_\Omega |D^\alpha u|^2 dx\Big)^{1/2}, \qquad (1)$$

where $\alpha = (\alpha_1,\alpha_2)$, $|\alpha| = \alpha_1 + \alpha_2$, $D^\alpha u = \partial^{|\alpha|} u/\partial x_1^{\alpha_1} \partial x_2^{\alpha_2}$.

By $H^2(\Omega)$ we denote the completion of $C^\infty(\bar{\Omega})$ (the space of infinitely differentiable function in $\bar{\Omega}$) in the norm (1). Set

$$\nabla u = \text{grad } u = (u_{x_1}, u_{x_2}),\ E(u,v) = \sum_{i,j=1}^{2} \frac{\partial^2 u}{\partial x_i \partial x_j} \frac{\partial^2 v}{\partial x_i \partial x_j},\ E(u) = E(u,u).$$

Let the point 0 be the origin and assume that $0 \in \partial\Omega$, where $\partial\Omega$ is the boundary of Ω. Set

$$\sigma_t = \partial\Omega \cap \{x: |x|=t\},\ S_t = \Omega \cap \{x: |x|=t\},\ \Omega_t = \Omega \cap \{x: |x|<t\}.$$

Suppose that the constant ω is such that $\pi \leq \omega \leq 2\pi$. Consider the equation

$$\sin^2(\omega\delta) = \delta^2 \sin^2\omega, \tag{2}$$

where δ is to be found. One can easily see that for $\pi \leq \omega \leq 2\pi$ there is a unique solution $\delta(\omega)$ of (2) such that $0 < \omega\delta(\omega) \leq \pi$. It is evident that $\delta(2\pi) = 1$. In what follows we shall assume that $1{,}24\pi \leq \omega \leq 2\pi$. One can show that in this case we have $1/2 \leq \delta(\omega) \leq 3/4$. For such values of ω the main results for the biharmonic equation with the right-hand side of the form

$$f + \sum_{j=1}^{2} \frac{\partial f_j}{\partial x_j},$$

where $f \in L^s(\Omega)$, $f_j \in L^p(\Omega)$, $j=1,2$, are proved in [2], [3] for $s \geq 1$ and some $p > 1$. Some of these results will be used in this paper.

2. In a bounded plane domain Ω we consider the equation

$$\Delta\Delta u = f + \sum_{j=1}^{2} \frac{\partial f_j}{\partial x_j} + \sum_{|\alpha|=2} D^\alpha f_\alpha \tag{3}$$

with the boundary conditions

$$u|_{\partial\Omega} = 0,\quad \frac{\partial u}{\partial \nu}\Big|_{\partial\Omega} = 0, \tag{4}$$

where ν is the outward normal to $\partial\Omega$, $f \in L^s(\Omega)$,

$$f_j \in L^p(\Omega),\ j=1,2,\ f_\alpha \in L^q(\Omega),\ |\alpha| = 2,\ s \geq 1,\ p > 1,\ q > 2.$$

The more precise condition on p and q will be given below.

A function $u(x) \in \overset{\circ}{H}{}^2(\Omega)$ is called a weak solution of problem (3), (4) if the integral identity.

$$\int_\Omega E(u,v)dx = \int_\Omega fvdx - \sum_{j=1}^{2} \int_\Omega f_j \frac{\partial v}{\partial x_j} dx + \sum_{|\alpha|=2} \int_\Omega f_\alpha D^\alpha v dx \qquad (5)$$

holds for any $v(x) \in \overset{\circ}{H}{}^2(\Omega)$.

It is easy to show that a weak solution of problem (3), (4) exists and it is unique for $f \in L^s(\Omega)$, $f_j \in L^p(\Omega)$, $j=1,2$, $f_\alpha \in L^q(\Omega)$, $|\alpha| = 2$, $s \geq 1$, $p > 1$, $q \geq 2$.

Lemma 1. *Let* $v \in \overset{\circ}{H}{}^1(\Omega)$. *Suppose that the length of the largest are belonging to* S_t *is not greater than* ωt, *the set* σ_t *is not empty if* S_t *is not empty for all* $t > 0$. *Then*

$$\int_\Omega |x|^\sigma |v|^\theta dx \leq C\left[\int_\Omega |x|^\gamma |\nabla v|^2 dx\right]^{\theta/2}, \qquad \theta > 0, \qquad (6)$$

where $\sigma > -2 + \gamma\theta/2$, *the constant* C *depends on* $\omega, \sigma, \theta, \gamma$ *and the diameter of* Ω *only.*

Proof. Consider the domain $\Omega(\rho,2\rho) = \Omega \cap \{x:\rho<|x|<2\rho\}$. It follows from the Sobolev imbedding theorem that for any $\theta > 0$

$$\left(\int_{\Omega(\rho,2\rho)} |v|^\theta dx\right)^{1/\theta} \leq C_1\left[\rho^{2/\theta}\left(\int_{\Omega(\rho,2\rho)} |\nabla v|^2 dx\right)^{1/2} + \right. \qquad (7)$$
$$\left. + \rho^{-1+2/\theta}\left(\int_{\Omega(\rho,2\rho)} |v|^2 dx\right)^{1/2}\right],$$

where the constant C_1 does not depend on ρ. This estimate is obtained from the corresponding imbedding theorem for the domain $\Omega(1,2)$ in the plan (x_1',x_2') by the transformation of the independent variables $x = \rho x'$.

From (7) we have

$$\rho^{-2/\theta+\gamma/2}\left(\int_{\Omega(\rho,2\rho)}|v|^{\theta}dx\right)^{1/\theta} \le C_1\left[\rho^{\gamma/2}\left(\int_{\Omega(\rho,2\rho)}|\nabla v|^2dx\right)^{1/2} + \right. \tag{8}$$

$$\left. + \rho^{-1+\gamma/2}\left(\int_{\Omega(\rho,2\rho)}v^2dx\right)^{1/2}\right]$$

By C_j we denote constants, independent of ρ and v. Since $\rho < |x| < 2\rho$ for $x \in \Omega(\rho,2\rho)$, it follows from (8) that

$$\left(\int_{\Omega(\rho,2\rho)}|x|^{-2+\gamma\theta/2}|v|^{\theta}dx\right)^{1/\theta} \le C_2\left[\left(\int_{\Omega(\rho,2\rho)}|x|^{\gamma}|\nabla v|^2dx\right)^{1/2} + \right. \tag{9}$$

$$\left. + \left(\int_{\Omega(\rho,2\rho)}|x|^{\gamma-2}|v|^2dx\right)^{1/2}\right].$$

To estimate the last integral we use lemma 1 from paper [3]. We then get from estimate (9) that

$$\left(\int_{\Omega(\rho,2\rho)}|x|^{-2+\gamma\theta/2}|v|^{\theta}dx\right)^{1/\theta} \le C_3\left[\int_{\Omega(\rho,2\rho)}|x|^{\gamma}|\nabla v|^2dx\right]^{1/2} \tag{10}$$

Let $\sigma > -2 + \gamma\theta/2$ and $\mu = \sigma - (-2+\gamma\theta/2)$, $\rho = 2^{-m-1}R$. It follows from (10) that

$$\int_{\Omega(2^{-m-1}R,2^{-m}R)}|x|^{\sigma}|v|^{\theta}dx \le C_4 2^{-m\mu}\left(\int_{\Omega(2^{-m-1}R,2^{-m}R)}|x|^{\gamma}|\nabla v|^2dx\right)^{\theta/2} \tag{11}$$

Summing up inequalities (11) with respect to m from 0 to ∞, we obtain (6).

Remark. Suppose that the set σ_t is non-empty if S_t is non-empty, for $0 < t \le t_1$, where t_1 is a positive constant. Then estimate (6) in Lemma 1 should be replaced by the following inequality

$$\int_{\Omega}|x|^{\sigma}|v|^{\theta}dx \le C\left[\left(\int_{\Omega}|x|^{\gamma}|\nabla v|^2dx\right)^{\theta/2} + \left(\int_{\Omega}v^2dx\right)^{\theta/2}\right],$$

where the constant C depends on $t_1,\omega,\gamma,\theta,\sigma$ and the diameter of Ω only.

Theorem 1. *Let* $1{,}24\pi \le \omega \le 2\pi$. *Suppose that the length of the largest arc belonging to* S_t *is not greater then* ωt *and* $\sigma_t \ne \emptyset$, *if* $S_t \ne \emptyset$. *Then the solution* $u(x)$ *of problem* (3), (4) *satisfies the inequalities.*

$$\int_{\Omega_t} E(u)dx \le Ct^{2\delta(\omega)}\Big[\|f\|^2_{L^s(\Omega)} + \sum_{j=1}^{2}\|f_j\|^2_{L^p(\Omega)} + \sum_{|\alpha|=2}\|f_\alpha\|^2_{L^q(\Omega)}\Big] \quad (12)$$

$$|u(x)|^2 \le C|x|^{2+2\delta(\omega)}\Big[\|f\|^2_{L^s(\Omega)} + \sum_{j=1}^{2}\|f_j\|^2_{L^p(\Omega)} + \sum_{|\alpha|=2}\|f_\alpha\|^2_{L^q(\Omega)}\Big] \quad (13)$$

where $\delta(\omega)$ *is a root of the equation* $\sin^2(\omega\delta) = \delta^2\sin^2\omega$ *such that* $0 < \omega\delta(\omega) \le \pi$, *the constant* C *depends only on* ω, p, q, s *and the diameter* R *of* Ω ; $s \ge 1$, $p > 2/(2-\delta(\omega))$, $q > 2/(1-\delta(\omega))$.

Proof. Consider a sequence of smooth domains Ω^m such that $\bar\Omega^m \subset \bar\Omega^{m+1} \subset \Omega$, $\bigcup_m \Omega^m = Q$. Let f^m, f_j^m, f_α^m be sequences of infinitely differentiable functions converging to f, f_j, f_α, $j=1,2$, $|\alpha| = 2$, in $L^s(\Omega)$, $L^p(\Omega)$, $L^q(\Omega)$ respectively. Denote by u_m the solution of the following problem

$$\Delta\Delta u_m = f^m + \sum_{j=1}^{2}\frac{\partial f_j^m}{\partial x_j} + \sum_{|\alpha|=2} D^\alpha f_\alpha^m \text{ in } \Omega^m, \quad (14)$$

$$u_m\big|_{\partial\Omega^m} = 0, \quad \frac{\partial u_m}{\partial\nu_m}\Big|_{\partial\Omega^m} = 0. \quad (15)$$

Set $\Phi(|x|) = |x|^{-2\delta(\omega)}$ for $|x| \ge t$, $\Phi(|x|) = t^{-2\delta(\omega)} + 2\delta(\omega)(t-|x|)t^{-2\delta(\omega)-1}$ for $|x| < t$, $t = \text{const} > 0$. Multipying both sides of (14) by $\Phi(|x|)u_m(x)$, integrating them over Ω^m, transforming them by intergation by parts and taking into account (15), we obtain

$$\mathfrak{J} \equiv \int_{\Omega^m} E(u_m,\Phi u_m)dx = \int_{\Omega^m} f^m\Phi u_m dx - \int_{\Omega^m}\sum_{j=1}^{2} f_j^m\frac{\partial}{\partial x_j}(\Phi u_m)dx +$$

$$+ \int_{\Omega^m}\sum_{|\alpha|=2} f_\alpha^m D^\alpha(\Phi u_m)dx \equiv \mathfrak{J}_1 + \mathfrak{J}_2 + \mathfrak{J}_3 .$$

It follows from inequalities (32), (33) proved in paper [2] that

$$\int_{\Omega_t \cap \Omega^m} E(u_m)dx \le C_1 t^{2\delta(\omega)} |\mathfrak{J}|, \tag{16}$$

where C_1 depends on ω only. By C_j we shall denote constants which do not depend on ρ and m. We assume that $u_m = 0$ outside of Ω^m. Let $\varepsilon = \text{const} > 0$, $Q_\rho = \{x:\rho<|x|<2\rho\}$. It is evident that

$$\begin{aligned}\int_{Q_\rho} |x|^{-2\delta(\omega)+\varepsilon} E(u_m)dx &\le C_2 \rho^{\varepsilon-2\delta(\omega)} \int_{Q_\rho} E(u_m)dx \\ &\le C_3 \rho^{\varepsilon} \sup_{0<t<\infty} \{t^{-2\delta(\omega)} \int_{\Omega_t} E(u_m)dx\}\end{aligned} \tag{17}$$

Setting $\rho = 2^{-m-1}R$, $m \ge 0$ and summing up inequalities (17) with respect to m from 0 to ∞, we get

$$\int_{\Omega} |x|^{-2\delta(\omega)+\varepsilon} E(u_m)dx \le C_4 A, \quad A = \sup_{0<t<\infty} \{t^{-2\delta(\omega)} \int_{\Omega_t} E(u_m)dx\}. \tag{18}$$

The imbedding theorem, lemma 1 from [3] and inequality (18) imply for $|x| = \rho$

$$\begin{aligned}|u(x)|^2 &\le C_5 \int_{Q_\rho} (\rho^{-2} u_m^2 + |\nabla u_m|^2 + E(u_m)\rho^2)dx \le \\ &\le C_6 \int_{Q_\rho} (|x|^{-2} u_m^2 + |\nabla u_m|^2 + |x|^2 E(u_m))dx \le C_7 \int_{Q} |x|^2 E(u_m)dx \le \\ &\le C_8 |x|^{2\delta(\omega)-2\varepsilon+2} \int_{Q_\rho} |x|^{-2\delta(\omega)+2\varepsilon} E(u_m)dx \le C_9 A |x|^{2\delta(\omega)-2\varepsilon+2}.\end{aligned} \tag{19}$$

Let us estimate $\mathfrak{J}$. Taking into account (19) and the fact that $\delta(\omega) \le 3/4$, we get

$$|\mathfrak{J}_1| = |\int_{\Omega^m} f^m \Phi u_m dx| \le C_{10} A^{1/2} \int_{\Omega^m} |f^m| |x|^{\delta(\omega)-\varepsilon+1} dx \le C_{11} A^{1/2} \|f^m\|_{L^s(\Omega^m)},$$

provided that $\varepsilon < 1 - \delta(\omega)$. It is evident that

$$| \mathfrak{J}_2| = |\int_{\Omega^m} \sum_{j=1}^{2} f_j^m \frac{\partial(\Phi u_m)}{\partial x_j} dx| \le \sum_{j=1}^{2} \int_{\Omega^m} |f_j^m||\Phi||\frac{\partial u_m}{\partial x_j}|dx + \tag{20}$$

$$+ \sum_{j=1}^{2} \int_{\Omega^m} |f_j^m||\frac{\partial \Phi}{\partial x_j}||u_m|dx.$$

By Hölder's inequality we obtain

$$\int_{\Omega^m} |f_j^m|\Phi|\frac{\partial u_m}{\partial x_j}|dx \le (\int_{\Omega^m} |f_j^m|^p dx)^{1/p} (\int_{\Omega^m} |x|^{-2\delta(\omega)p'} |\frac{\partial u_m}{\partial x_j}|^{p'} dx)^{1/p'} \le$$

$$\le C_{12} \|f\|_{L^p(\Omega^m)} (\int_{\Omega^m} E(u^m)|x|^\gamma dx)^{\theta/2p'}, \quad 1/p + 1/p' = 1.$$

Here we have applied lemma 1 to the function $\frac{\partial u_m}{\partial x_j}$, setting $\theta = p'$, $\sigma = -2\delta(\omega)p'$, $\gamma = -2\delta(\omega) + \varepsilon$, since $\sigma > -2 + \gamma\theta/2$. The last inequality holds for sufficiently small ε and $p > 2/(2-\delta(\omega))$. Therefore,

$$\int_{\Omega^m} |f_j^m||\frac{\partial u_m}{\partial x_j}|\Phi dx \le C_{13} \|f_j^m\|_{L^p(\Omega^m)} A^{1/2}.$$

In order to estimate the last integrals in (20), we use (19) and Hölder's inequality. We then get

$$\int_{\Omega^m} |f_j|^m \frac{\partial \Phi}{\partial x_j}||u_m|dx \le C_{14} \int_{\Omega^m} |f_j^m||x|^{-2\delta(\omega)-1} A^{1/2} |x|^{1+\delta(\omega)-\varepsilon} \le$$

$$\le C_{14} A^{1/2} \|f_j^m\|_{L^p(\Omega^m)} (\int_{\Omega^m} |x|^{-(\delta(\omega)+\varepsilon)p'} dx)^{1/p'} \le C_{15} A^{1/2} \|f_j^m\|_{L^p(\Omega^m)},$$

since $-(\delta(\omega)+\varepsilon)p' > -2$ for sufficently small ε and $p > 2/(2-\delta(\omega))$. Let us estimate $\mathfrak{J}_3$. It is easy to see that

$$|\int_{\Omega^m} f_\alpha^m D^\alpha(\Phi u_m)dx| \le \int_{\Omega^m} |f_\alpha^m|| D^\alpha \Phi||u_m|dx + \tag{21}$$

$$+ 2 \int_{\Omega^m} |f_\alpha^m||\nabla\Phi||u_m|dx + \int_{\Omega^m} |f_\alpha^m||\Phi||D^\alpha u_m|dx = \mathfrak{J}_4 + \mathfrak{J}_5 + \mathfrak{J}_6.$$

Using inequality (19), we obtain

$$\mathfrak{J}_4 \le C_{16}A^{1/2}\int_{\Omega^m}|f_\alpha^m||x|^{-2\delta(\omega)-2}|x|^{1+\delta(\omega)-\varepsilon}dx \le$$

$$\le C_{16}A^{1/2}\|f_\alpha^m\|_{L^q(\Omega^m)}\left(\int_{\Omega^m}|x|^{-(\delta(\omega)+1+\varepsilon)q'}\right)^{1/q'} \le C_{17}A^{1/2}\|f_\alpha^m\|_{L^q(\Omega^m)},\ 1/q+1/q'=1,$$

since $-(\delta(\omega)+1+\varepsilon)q' > -2$ for sufficiently small ε and $q > 2/(1-\delta(\omega))$. Taking into account Hölder's inequality, we find that

$$\mathfrak{J}_5 \le C_{18}\|f_\alpha^m\|_{L^q(\Omega^m)}\left(\int_{\Omega^m}|x|^{-(2\delta(\omega)+1)q'}|\nabla u_m|^{q'}dx\right)^{1/q'}. \qquad (22)$$

To estimate the last integral we use lemma 1 for functions $\frac{\partial u_m}{\partial x_j}$, j=1,2, with $\theta = q'$, $\sigma = -(2\delta(\omega)+1)q'$, $\gamma = -2\delta(\omega) + \varepsilon$. The condition $\sigma > -2 + \gamma\theta/2$ is satisfied for small ε and $q > 2/(1-\delta(\omega))$. We have

$$\int_{\Omega^m}|x|^{-(2\delta(\omega)+1)q'}|\nabla u_m|^{q'}dx \le C_{19}\int_{\Omega^m}|x|^{-2\delta(\omega)+\varepsilon}E(u_m)dx. \qquad (23)$$

It follows from (22), (23) that

$$\mathfrak{J}_5 \le C_{20}\|f_\alpha^m\|_{L^q(\Omega^m)}A^{1/2}.$$

It is easy to see that

$$\mathfrak{J}_6 \le \left(\int_{\Omega^m}|x|^{-2\delta(\omega)+\varepsilon}|D^\alpha u_m|^2dx\right)^{1/2}\left(\int_{\Omega^m}|f_\alpha^m|^2|x|^{-2\delta(\omega)-\varepsilon}dx\right)^{1/2} \le$$

$$\le C_{21}A^{1/2}\left(\int_{\Omega^m}|f_\alpha^m|^q dx\right)^{1/q}\left(\int_{\Omega^m}|x|^{-(2\delta(\omega)+\varepsilon)q/(q-2)}dx\right) \le C_{22}A^{1/2}\|f^m\|_{L^q(\Omega^m)},$$

since the condition $q > 2/(1-\delta(\omega))$ implies that $-(2\delta(\omega)+\varepsilon)q/(q-2) > -2$ for small ε.

Therefore, the estimate obtained for $|\mathfrak{J}|$ and inequality (16) yield

$$t^{-2\delta(\omega)}\int_{\Omega_t} E(u_m)dx \le C_{23}A^{1/2}\Big(\|f^m\|_{L^s(\Omega^m)} + \sum_{j=1}^{2}\|f_j^m\|_{L^p(\Omega^m)} + \sum_{|\alpha|=2}\|f_\alpha^m\|_{L^q(\Omega^m)}\Big).$$

It follows that

$$t^{-2\delta(\omega)}\int_{\Omega_t} E(u_m)dx \le C_{24}\Big[\|f^m\|_{L^s(\Omega^m)} + \sum_{j=1}^{2}\|f_j^m\|_{L^p(\Omega^m)} + \sum_{|\alpha|=2}\|f_\alpha^m\|_{L^q(\Omega^m)}\Big].$$

Making m tend to infinity, we obtain from the last inequality estimate (12). Estimate (13) follows directly from (12). Indeed, applying the imbedding theorem and lemma 1 from [3] we get

$$|u(x)|^2 \le C\rho^2 \int_{\Omega(\rho,2\rho)} E(u)dx \text{ for } x \in \Omega(\rho,2\rho). \tag{24}$$

Estimating the last integral by (12), we obtain (13) from (24). The theorem is proved.

In order to estimate the first derivatives of a weak solution $u(x)$ of problem (3), (4) in Ω, we need the following.

Lemma 2. *Let* $u_1(x)$ *be a weak solution of equation* (3) *with* $f \equiv 0$, $f_\alpha \equiv 0$, $|\alpha| = 2$, *in the circle* $K_\rho = \{x: |x-x^\circ| < \rho\}$ *and let* $u_2(x)$ *be a weak solution of equation* (3) *with* $f \equiv 0$, $f_j \equiv 0$, $j=1,2$, *in* K_ρ. *Then for* $x \in K_{\rho/2} = \{x: |x-x^\circ| < \rho/2\}$

$$|\nabla u_1(x)| + |\nabla u_2(x)| \le C \sum_{|\alpha|=2}\|f_\alpha\|_{L^q(K_\rho)}\rho^{1-2/q} + \sum_{j=1}^{2}\|f_j\|_{L^p(K_\rho)}\rho^{2-2/p} + \sum_{j=1}^{2}\|u_j\|_{L^2(K_\rho)}\rho^{-2}, \tag{25}$$

where the constant C *does not depend on* ρ.

Proof. It follows from the results in the theory of elliptic equations (see [4]) that

$$\|u_1\|_{W_p^3(K_{1/2})} \le C_1[\sum_{j=1}^{2} \|f_j\|_{L^p(K_1)} + \|u_1\|_{L^2(K_1)}], \ p > 1, \qquad (26)$$

$$\|u_2\|_{W_q^2(K_{1/2})} \le C_2[\sum_{|\alpha|=2} \|f_\alpha\|_{L^q(K_1)} + \|u_2\|_{L^2(K_1)}], \ q > 1, \qquad (27)$$

where C_1, C_2 = const, u_1, u_2 are weak solutions of the corresponding equations in K_ρ for $\rho=1$, W_p^ℓ is the Sobolev space formed by function which belongs to $L^p(K)$ and their derivatives up to the order ℓ belong to $L^p(K)$.

The Sobolev imbedding theorems and estimates (26), (27) imply that for $x \in K_{1/2}$ the following inequalities are valid :

$$|\nabla u_1(x)| \le C_3\|u\|_{W_p^3(K_{1/2})} \le C_4(\sum_{j=1}^{2} \|f_j\|_{L^p(K_1)} + \|u_1\|_{L^2(K_1)}), \qquad (29)$$

$$|\nabla u_2(x)| \le C_5\|u\|_{W_q^2(K_{1/2})} \le C_6(\sum_{|\alpha|=2} \|f_\alpha\|_{L^q(K_1)} + \|u_2\|_{L^2(K_1)}). \qquad (29)$$

From (28), (29) we obtain estimate (25) for $\rho=1$. In order to get (25) for any $\rho > 0$, we introduce new variables $x' = (x-x^\circ)/\rho$ in equation (3). Then K_ρ is mapped to K_1 and equation (3) takes the form

$$\Delta\Delta u' = \rho^4 f + \sum_{j=1}^{2} \rho^3 \frac{\partial f_j}{\partial x'_j} + \rho^2 \sum_{|\alpha|=2} D^\alpha_{x'} f_\alpha. \qquad (30)$$

Inequality (25) holds in K_1 for the function $u_1'(x') = u_1(x)$ and $u_2'(x') = u_2(x)$, which satisfy the corresponding equations of the form (30). These inequalities in variables x give (25) for any ρ. The lemma is proved.

Theorem 2. *Suppose that all conditions of theorem 1 are satisfied. Moreover, let the circle* $|x-x^\circ| = \rho$ *for* $\rho \le |x^\circ|/2$ *have a non-empty intersection with* $\partial\Omega$ *for all* $x^\circ \in \partial\Omega$ *such that* $|x^\circ| < \rho_1$, ρ_1 = const > 0. *Then for* $|x| < \rho_1/2$ *the inequality*

$$|\nabla u(x)|^2 \le C|x|^{2\delta(\omega)}\{\|f\|^2_{L^s(\Omega)} + \sum_{j=1}^{2}\|f_j\|^2_{L^p(\Omega)} + \sum_{|\alpha|=2}\|f_\alpha\|_{L^q(\Omega)}\} \quad (31)$$

holds with a constant C depending only on ω, p, q, s and the diameter of Ω.

Proof. Estimate (31) is proved in paper [3] (theorem 4) for weak solutions of problem (3), (4) with $f \in L^s(\Omega)$, $s \ge 1$ and $f_j \equiv 0$, $j=1,2$, $f_\alpha \equiv 0$, $|\alpha| = 2$. Therefore, it is sufficient to prove (31) for $f \equiv 0$ in (3). This can be done by the use of lemma 2 and estimates (12), (13) in the same way as in theorem 5 of paper [3].

Remark. The condition of theorem 1 and 2 that $\sigma_t \ne \emptyset$, if $S_t \ne \emptyset$, for all $t < \infty$ can be replaced by the condition that $\sigma_t \ne \emptyset$, if $S_t \ne \emptyset$, for $t \le t_0 < \infty$, $t_0 = \text{const} > 0$, and boundary condition (4) can be replaced by the condition.

$$u|_\Gamma = \frac{\partial u}{\partial \nu}\Big|_\Gamma = 0,$$

where $\Gamma \subset \partial\Omega$, $0 \in \Gamma$, $\partial\Omega_{t_0} \cap \partial\Omega \subset \Gamma$. Then inequalities similar to (12), (13) (31) hold for $u(x)$ provided that other conditions of theorem 1 and 2 are satisfied. They can be proved, using theorem 1.2, in the same way as theorems 3,4 are obtained in paper [3].

Theorem 3. *Suppose that for all* $x^\circ \in \partial\Omega$ *the intersection of* $\partial\Omega$ *with the circle* $|x-x^\circ| = \rho$ *for* $\rho \le t_0$ *is non-empty and the length of the largest arc belonging to* $\Omega \cap \{x: |x-x^\circ| = \rho\}$ *is not greater than* $\omega\rho$, *and*

$$f \in L^s(\Omega),\ f_j \in L^p(\Omega),\ j=1,2,\ f_\alpha \in L^q(\Omega),\ |\alpha| = 2,$$
$$s \ge 1,\ p > 2/(2-\delta(\omega)),\quad q > 2/(1-\delta(\omega)).$$

Then the function $u(x)$ *which is a weak solution of the problem belongs to the class* $C^{1+\delta(\omega)}(\bar\Omega)$, *i.e. the first derivatives of* $u(x)$ *satisfy the Hölder condition in* $\bar\Omega$ *with the exponent* $\delta(\omega)$.

The proof of this theorem is completely identical to that of theorem 5 in paper [3].

It is shown in paper [3] that in some sense theorem 1.3 can not be improved, namely the number $\delta(\omega)$ can not be replaced by $\delta(\omega) + \varepsilon$ with $\varepsilon > 0$. These theorems are not valid, if $p < 2/(2-\delta(\omega))$ or $q < 2/(1-\delta(\omega))$. In order to show this let us consider the equation

$$\Delta\Delta u = \frac{\partial r^\lambda}{\partial x_1}, \tag{32}$$

where $r = |x|$, $\lambda = -2/p+\varepsilon$, $p < 2/(2-\delta(\omega))$, $\varepsilon = \text{const} > 0$. For any $\varepsilon > 0$ the function $r^\lambda \in L^p(\Omega)$. Set $p = 2/(2-\delta(\omega))(1+\varepsilon_i)$, where $\varepsilon_1 = \text{const} > 0$. It is easy to see that in the domain $\Omega = \{x : r < 1,\ 0 < \zeta < \omega\}$, (r,ζ) are polar coordinates in (x_1,x_2) - plane, equation (32) admits a solution of the form $v(x) = r^{\lambda+3}\Phi(\zeta)$ for a sequence of ε, tending to zero, and also $v \in H^2(\Omega)$, $u = \frac{\partial u}{\partial \nu} = 0$ for $\zeta = 0$ and $\zeta = \omega$. We have

$$\frac{\partial v}{\partial r} = (\lambda+3)r^{\lambda+2}\Phi(\zeta),\ \lambda + 2 = \delta(\omega) - \varepsilon_1\delta(\omega) + \varepsilon - 2\varepsilon_1 < \delta(\omega)$$

for any small ε_1, if ε is sufficiently small, since $\delta(\omega) < 1$. The function $u(x) = v(x)\psi(x)$ where $\psi(x) \equiv 1$ for $r < 1/2$ and $\psi \equiv 0$ for $r > 3/4$, $\psi \in C(\bar{\Omega})$, is a solution of problem (3), (4) with $f \in L^s(\Omega)$, $s \geq 1$, $f_j \in L^p(\Omega)$, $p < 2/(2-\delta(\omega))$, $f_\alpha \equiv 0$, $|\alpha| = 2$, and $u(x)$ does not belong to the class $C^{1+\delta}(\omega)$ and its gradient does not satisfy estimate (31).

In a similar way one can construct a function $v(x) = r^{\lambda+2}\Phi_1(\zeta)$, such that $\lambda = -2/q + \varepsilon$, $q = 2/(1-\delta(\omega))(1+\varepsilon_1) < 2/(1-\delta(\omega))$, $\varepsilon_1 = \text{const} > 0$, $\Delta\Delta v = \frac{\partial^2 r}{\partial x_1^2}$, $v|_{\zeta=0} = \frac{\partial v}{\partial \nu}|_{\zeta=\omega} = \frac{\partial v}{\partial \nu}|_{\zeta=\omega} = 0$. The function $u = v\psi \in \overset{\circ}{H}{}^2(\Omega)$ and it does not satisfy (31) in a neignbourhood of $x = 0$.

3. Let us consider a nonhomogeneous Dirichlet boundary value problem for equation (3) in Ω with the boundary conditions

$$u = \Phi,\ \frac{\partial u}{\partial v} = \frac{\partial \Phi}{\partial \nu} \text{ on } \Omega, \tag{33}$$

where $\Phi \in W_q^2(\Omega)$, $q > 2/(1-\delta(\omega))$.

We say that u(x) is a weak solution of the problem (3), (33), if $u \in H^2(\Omega)$, $w = u - \Phi \in \overset{\circ}{H}{}^2(\Omega)$ and the integral identity

$$\int_\Omega E(w,v)dx = \int_\Omega (fv - \sum_{j=1}^{2} f_j \frac{\partial v}{\partial x_j} + \sum_{|\alpha|=2} f_\alpha D^\alpha v - \sum_{j=1}^{2} \frac{\partial^2 \Phi}{\partial x_i \partial x_j} \frac{\partial^2 v}{\partial x_i \partial x_j})dx$$

holds for any $v \in \overset{\circ}{H}{}^2(\Omega)$.

Since $\Phi \in W_q^2(\Omega)$, $q > 2/(1-\delta(\omega))$, $f \in L^s(\Omega)$, $s \geq 1$, $f_j \in L^p(\Omega)$, $p > 2/(2-\delta(\omega))$, j=1,2, $f_\alpha \in L^q(\Omega)$, $q > 2/(1-\delta(\omega))$, for w theorems 1-3 are valid. According to the imbedding theorems (see [5], p. 93, [6], p. 440), we have

$$\Phi \in C^{1+\beta}(\bar{\Omega}), \ \beta < (q-2)/q,$$

provided that the domain Ω is either starshaped with respect to a convex subdomain of Ω or Ω is a finite sum of domains of this kind. In particular, one can take $\beta = \delta(\omega)$.

Thus, for the weak solution $u = w + \Phi$ of problem (3), (33) theorem 3 is valid, which means that under conditions of theorem 3 imposed on Ω and functions f, f_j, f_α the weak solution belongs to $C^{1+\delta(\omega)}$, if $\Phi \in W_q^2$, $q > 2/(1-\delta(\omega))$ and Ω has properties, mentioned above.

4. Let us consider the case when the boundary $\partial\Omega$ in a neighbourhood of 0 is such that the set $\sigma_t = \partial\Omega \cap \{x:|x|=t\}$ can be empty for anyhow small values of t. As before we assume that the origin 0 belongs to $\partial\Omega$.

Lemma 3. *Suppose that* $u \in H^2(\Omega)$ *and there exists three points* $P_1 = (x_{11},x_{21})$, $P_2 = (x_{12},x_{22})$, $P_3 = (x_{13},x_{23})$ *of the domain* G *which do not belong to a straight line and such that*

$$u(P_j) = 0, \quad j=1,2,3.$$

Suppose that G *is either starshaped with respect to a convex subdomain of* Ω *or is a finite sum of domains of this kind. Then*

$$\int_G u^2 dx + \int_G |\nabla u|^2 dx \le C\, B^{-2} \int_G \sum_{|\alpha|=2} |D^\alpha u|^2 dx,$$

where constant C *does not depend on* $u(x)$ *and the position of points* P_j, $j=1,2,3$;

$$B = \begin{vmatrix} 1 & x_{11} & x_{21} \\ 1 & x_{12} & x_{22} \\ 1 & x_{13} & x_{23} \end{vmatrix}$$

Proof. By $\bar{v}$ we denote the mean value of a function v in G, i.e.

$$\bar{v} = (\text{mes } G)^{-1} \int_G v\, dx.$$

Consider the function

$$v(x) = u(x) - \mu - \bar{u}_1 x_1 - \bar{u}_2 x_2,$$

where $\bar{u}_j$ is a mean value of $\frac{\partial u}{\partial x_j}$ in G, $j=1,2$; $\mu = \bar{u} - \bar{u}_1\bar{x}_1 - \bar{u}_2\bar{x}_2$. It is easy to see that $\bar{v} = \bar{v}_1 = \bar{v}_2 = 0$. Therefore, applying the Poincare inequality (see [6], p. 436) to v and its derivatives $\frac{\partial v}{\partial x_1}$, $\frac{\partial v}{\partial x_2}$ we get

$$\int_G v^2 dx + \int_G |\nabla v|^2 dx \le C_1 \int_G \sum_{|\alpha|=2} |D^\alpha v|^2 dx, \tag{35}$$

where the constant C_1 depends on the domain G only. It follows that

$$\int_G (u^2 + |\nabla u|^2) dx \le C_2 \left(\int_G \sum_{|\alpha|=2} |D^\alpha u|^2 dx + |\bar{u}|^2 + |\bar{u}_1|^2 + |\bar{u}_2|^2 \right). \tag{36}$$

Let us estimate $\bar{u}_1$, $\bar{u}_2$, $\bar{u}$. Due to the assumption of the theorem $u(P_j) = 0$, $j=1,2,3$. Hence, we have

$$v(P_i) = -\mu - \bar{u}_1 x_{1i} - \bar{u}_2 x_{2i}, \quad i=1,2,3 \tag{37}$$

Taking into account the formula for μ, we obtain from (37) the system of equations for $\bar{u}$, $\bar{u}_1$, $\bar{u}_2$ of the form

$$- \bar{u} + \bar{u}_1\bar{x}_1 + \bar{u}_2\bar{x}_2 - \bar{u}_1 x_{1i} - \bar{u}_2 x_{2i} = v(P_i), \quad i=1,2,3. \tag{38}$$

It is easy to see that the determinant of this system is equal to $- B \neq 0$. We find from (38) that

$$|\bar{u}| + |\bar{u}_1| + |\bar{u}_2| \le C_3|B|^{-1}(|v(P_1)| + |v(P_2)| + |v(P_3)|), \tag{39}$$

where C_3 depends on the diameter of G only. By virtue of the imbedding theorem [5] we have

$$|v(P)| \le C_4\|v\|_2, \quad P \in G \tag{40}$$

It follows from estimates (40), (35), (39) that

$$|\bar{u}| + |\bar{u}_1| + |\bar{u}_2| \le C_5|B|^{-1}\Big(\sum_{|\alpha|=2} |D^\alpha u|^2 dx\Big)^{1/2} \tag{41}$$

Estimates (36) and (41) yield

$$\int_G u^2 dx + \int_G |\nabla u|^2 dx \le C_6(1+|B|^{-2}) \int_G \sum_{|\alpha|=2} |D^\alpha u|^2 dx.$$

Since B is proportional to the area of the triangle $\Delta P_1P_2P_3$ and the domain G is bounded, then $|B|^{-1} > C_7$ with C_7 depending on G only. Therefore inequality (34) is valid. The lemma is proved.

Lemma 4. *Let* $u \in H^2(Q_\rho)$, *where* $Q_\rho = \{x:\rho<|x|<2\rho\}$. *Suppose that there are three points* $P_1 = (x_{11},x_{12})$, $P_2 = (x_{12},x_{22})$, $P_3 = (x_{13},x_{23})$ *in* Q_ρ *such that* $u(P_j) = 0$, $j=1,2,3$.

Then

$$\int_{Q_\rho} \rho^{-4}u^2 dx + \int_{Q_\rho} \rho^{-2}|\nabla u|^2 dx \le C|B|^{-2} \int_{Q_\rho} \rho^4 \sum_{|\alpha|=2} |D^\alpha u|^2 dx, \tag{42}$$

where constant C *does not depend on* u *and* ρ.

Proof. We introduce new variables $x' = x/\rho$. Then in Q_1 in the variables x' estimate (34) holds for function $u'(x') = u(x)$. Returning in this inequality to the variables x, we get (42).

Theorem 4. *Let* $u(x)$ *belong to* $\overset{\circ}{H}{}^2(\Omega)$ *and be a weak solution of problem* (3), (4) *in* Ω *with* $f_\alpha \equiv 0$, $|\alpha| = 2$, $f \in L^2(\Omega)$, $f_j \in L^2(\Omega)$, $j=1,2$. *Let* 0 *be the origin and* $0 \in \partial\Omega$. *Suppose that in every* $Q^k = \{x: 2^{-k} < |x| < 2^{-k+1}\}$, *if* $k \geq k°$, *there exist three points* $P_{1k} = (x_{11}^k, x_{21}^k)$, $P_{2k} = (x_{12}^k, x_{22}^k)$, $P_{3k} = (x_{13}^k, x_{23}^k)$, *such that* P_{1k}, P_{2k}, P_{3k} *do not belong to* Ω *and*

$$B_k = \begin{vmatrix} 1 & x_{11}^k & x_{21}^k \\ 1 & x_{12}^k & x_{22}^k \\ 1 & x_{13}^k & x_{23}^k \end{vmatrix} \geq M2^{-2k}, \tag{43}$$

where the constant M *does not depend on* k. *Then there is a constant* β, *depending on* M *and* Ω *and such that*

$$\int_\Omega |x|^{-\beta} \sum_{|\alpha|=2} |D^\alpha u|^2 dx + \int_\Omega |x|^{-\beta-2} |\nabla u|^2 dx + \\ + \int_\Omega |x|^{-\beta-4} u^2 dx \leq C[\int_\Omega f^2 dx + \int_\Omega \sum_{j=1}^2 f_j^2 dx], \tag{44}$$

$$|u(x)|^2 \leq C|x|^{2+\beta}[\int_\Omega (f^2 + \sum_{j=1}^2 f_j^2) dx],$$

where C *is a constant independent of* u.

Proof. Consider a sequence of smooth subdomains Ω^m such that $\bar{\Omega}^m \quad \Omega^{m+1}$, $\underset{m}{\cup}\Omega^m = \Omega$ and approximate the function f and f_j, $j=1,2$, in $L^2(\Omega)$ by sequences of infinitely differentiable functions f^m and f_j^m respectively. It is obvious that condition (43) still holds for Ω^m. Denote by u_m a weak solution of the following boundary value problem in Ω^m

$$\Delta\Delta u_m = f^m + \sum_{j=1}^2 \frac{\partial f_j^m}{\partial x_j} \text{ in } \Omega^m, \quad u_m = \frac{\partial u_m}{\partial \nu_m} = 0 \text{ on } \partial\Omega^m, \tag{45}$$

Functions u_m satisfy the integral identity

$$\int_{\Omega^m} E(u_m,v)dx = \int_{\Omega^m} fv\, dx - \int_{\Omega^m} \sum_{j=1}^{2} f_j \frac{\partial v}{\partial x_j} dx \tag{46}$$

for any $v \in \overset{\circ}{H}{}^2(\Omega)$. We take $v = u_m|x|^{-\beta}$ in (46), where β is a positive constant to be chosen later. It is easy to see that

$$E(u_m,u_m r^{-\beta}) = r^{-\beta}E(u_m,u_m) + \beta \sum_{\substack{|\alpha|=2,\\ |\gamma|=1}} D^{\alpha}u_m D^{\alpha}u_m \Phi_{\alpha\gamma}(x) + \tag{47}$$
$$+ \beta \sum_{|\alpha|=2} r^{\beta-2} u_m D^{\alpha}u_m \Phi_{\alpha}(x), \quad r = |x|,$$

where $\Phi_{\alpha\gamma}$, Φ_{α} are bounded functions when β belongs to any finite interval. Integrating (47) over Ω^m and using the inequality $2ab \geq -\varepsilon a^2 - b^2/\varepsilon$, $\varepsilon = \text{const} > 0$ we obtain

$$\int_{\Omega^m} E(u_m,r^{-\beta}u_m)dx \geq \int_{\Omega^m} r^{-\beta}E(u_m,u_m)dx -$$
$$- C_1\beta \int_{\Omega^m} r^{-\beta}E(u_m,u_m)dx - C_2 \int_{\Omega^m} r^{-\beta-2}|\nabla u_m|^2 dx - C_3\beta \int_{\Omega^m} r^{-\beta-4}|u_m|^2 dx, \tag{48}$$

where constants C_1,C_2,C_3 do not depend on m. We continue the function u_m to R^2 setting $u_m = 0$ outside of Ω^m. Using lemma 4 to estimate the last two integrals we get

$$\int_{\Omega^m} r^{-\beta-2}|\nabla u_m|^2 dx + \int_{\Omega^m} r^{-\beta-4}u_m^2 dx = \sum_{k=k^\circ} \int_{Q^k} (r^{-\beta-2}|\nabla u_m|^2 +$$
$$+ r^{-\beta-4}u_m^2)dx + \int_{2^{-k_0+1}<|x|<R} (r^{-\beta-2}|\nabla u_m|^2 + r^{-\beta-4}|u_m|^2)dx \tag{49}$$

Here R is a constant such that $\bar{\Omega} \subset \{x: |x|<R\}$. In order to estimate the sum on the right-hand side of (49), we use condition (43). Taking into account (43) we deduce from lemma 4 and estimate (42) that

$$\int_{Q^k} (2^{-k})^{-4-\beta} u_m^2 dx + \int_{Q^k} (2^{-k})^{-2-\beta} |\nabla u_m|^2 dx \le$$

$$\le C_4 |B_k|^{-2} \int_{Q^k} (2^{-k})^{4-\beta} \sum_{|\alpha|=2} |D^\alpha u_m|^2 dx \le C_5 \int_{Q^k} (2^{-k})^{-\beta} \sum_{|\alpha|=2} |D^\alpha u_m|^2 dx. \tag{50}$$

Here and in that follows constants C_j do not depend on m and k. Inequalities (50) imply that

$$\int_{Q^k} r^{-4-\beta} u_m^2 dx \le \int_{Q^k} r^{-2-\beta} |\nabla u_m|^2 dx \le C_6 (M^{-2}+1) \int_{Q^k} r^{-\beta} \sum_{|\alpha|=2} |D^\alpha u_m|^2 dx. \tag{51}$$

Applying lemma 1 to the last integral in (49), we get by virtue of (51)

$$\int_{\Omega^m} r^{-\beta-4} u_m^2 dx + \int_{\Omega^m} r^{-2-\beta} |\nabla u_m|^2 dx \le C_7 \int_{\Omega^m} r^{-\beta} \sum_{|\alpha|=2} |D^\alpha u|^2 dx. \tag{52}$$

If β is sufficiently small (β depends on M and the diameter of Ω), then we have from (48) that

$$\int_{\Omega^m} E(u_m, r^{-\beta} u_m) dx \ge \frac{1}{2} \int_{\Omega^m} r^{-\beta} E(u_m, u_m) dx. \tag{53}$$

It follows from (46) for $v = r^{-\beta} u_m$ and estimate (53) that

$$\int_{\Omega^m} r^{-\beta} E(u_m) dx \le 2 \left| \int_{\Omega^m} f r^{-\beta} u_m dx - \int_{\Omega^m} \sum_{j=1}^{2} f_j \frac{\partial (r^{-\beta} u_m)}{\partial x_j} dx \right|. \tag{54}$$

Using the inequality $2ab \le \varepsilon a^2 + b^2/\varepsilon$, we obtain from (54) that for any ε

$$\int_{\Omega^m} r^{-\beta} E(u^m) dx \le \varepsilon \int_{\Omega^m} r^{-\beta-4} u_m^2 dx + C(\varepsilon) \int_{\Omega^m} r^{4-\beta} f^2 dx +$$

$$+ \int_{\Omega^m} r^{-\beta-2} |\nabla u_m|^2 dx + C(\varepsilon) \int_{\Omega^m} \sum_{j=1}^{2} r^{2-\beta} f_j^2 dx. \tag{55}$$

From (54), (55), choosing a sufficiently small ε, we get estimate (44) for u_m. Passing to the limit in (44) for Ω^m as $m \to \infty$ we obtain estimate (44) for $u(x)$ since $u_m \to u$ in $H^2(\Omega)$ as $m \to \infty$. Let us prove inequality (45). We set $u(x) = 0$ outside of Ω. Due to imbedding theorems we have for $x \in Q_\rho$, $Q_\rho = \{x: \rho < x < 2\rho\}$, $|u(x)|^2 \le C_8[\rho^2 \int_{Q_\rho} E(u)dx + \int_{Q_\rho} |\nabla u|^2 dx + \rho^{-2} \int_{Q_\rho} u^2 dx]$,

where C_8 does not depend on ρ. By virtue of (42) we get

$|u(x)|^2 \le C_9[\rho^2 \int_{Q_\rho} E(u)dx + \rho^6 |B_k|^{-2} \int_{Q_\rho} E(u)dx]$ for $\rho = 2^{-k}$, $k \ge k°$, the constant C_9 does not depend on ρ. Since $B_k^{-2} \le M^{-2} \rho^{-4}$ for $\rho = 2^{-k}$, we have

$$|u(x)|^2 \le C_{10}\rho^2 \int_{Q_\rho} E(u)dx, \quad x \in Q_\rho. \tag{56}$$

It follows from (56) that

$$\rho^{-\beta-2}|u(x)|^2 \le C_{11}\rho^{-\beta}\int_{Q_\rho} E(u)dx \le C_{12} \int_{Q_\rho} |x|^{-\beta}E(u)dx \le C_{13}[\int_\Omega f^2 dx +$$

$$+ \int_\Omega \sum_{j=1}^{2} f_j^2 dx], \ x \in Q_p.$$

Therefore, estimate (45) is valid with a constant C independent of u. The theorem is proved.

REFERENCES.

[1] O.A. Oleinik, B.A. Kondratiev, I. Kopacek. Asymptotic properties of solutions of the biharmonic equation. Differ. uravn., 1981, v. 17, N°10, 1886-1899.

[2] V.A. Kondratiev, I. Kopacek, D.M. Lekveishvili, O.A. Oleinik. Unimprovable estimates in the Hölder Spaces and the exact Saint-Venant's principle for the biharmonic equation. Trudy Matem. Inst. AN SSSR, 1983, N°166, 91-106.

[3] V.A. Kondratiev, O.A. Oleinik. Unimprovable estimates in the Hölder spaces for weak solutions of the biharmonic equation, the Navier-Stokes system and the von Karman system in nonsmooth two-dimensional domains. Vestnik Mosk. Univer., ser. I, Matem. and Mech., 1983, N°6, 22-39.

[4] S. Agmon, A. Douglis, L. Nirenberg. Estimates near the boundary for solutions of elliptic partial differential equations satisfying general boundary conditions. Comm. on pure and appl. mathem., 1959, v.XII, 623-727.

[5] S.L. Sobolev. Some applications of the functional analysis to the mathematical physics. Novosibirsk, 1962.

[6] L.V. Kantorovich, G.P. Akilov. Functional analysis. Mosc., 1977.

V.A. KONDRATIEV & O.A. OLEINIK
Université of Moscow
M.G.U. 78
"K" Kwartal 133
MOSCOW 117234
SOVIET UNION

P KREE

New boundary value problems connected with multivalued stochastic differential equations

Let S be a state-space. For any process $\xi = (\xi_t)$ associated with an S-valued and time-homogeneous Markoff process, the law (or distribution) of ξ_t at a given time t is denoted Law $\xi_t = m_t = m_t(dx)$ in general and $p(t,y,dx)$ if the initial value of (ξ_t) is $\xi_o = y \in S$; p is called the transition probability. The corresponding positive measure $m_t(dx)dt$ on $S \times \mathbb{R}$ in denoted M where M vanishes for $t < 0$. For example $M = m(dx)dt = m \otimes dt$ is a tensor product if the process ξ is stationary. Considering Markoff processes associated with multivalued Stochastic Differential Equations (SDE) [3] we generally are interested in the characterization of M as a solution of some Boundary value Problem (BUP) and in the study of these problems, including smoothness ard description of singularities. Let us consider Ito's equation

$$\text{SDE} \qquad d\xi(t) = b(\xi(t))dt + \sigma(\xi(t))dW(t)$$

reflected by the boundary $\sum$ of a closed regular subset G of $\mathbb{R}^n$. The function $b : \mathbb{R}^n \to \mathbb{R}^n$ and $\sigma : \mathbb{R}^n \to \text{Mat}\ (n,n)$ are C^∞, with linear-growth s.t. all their derivatives of b and σ are bounded. The transpose $\mathfrak{L}^T$ of the diffusion operator $\mathfrak{L}$ of SDE :

$$\mathfrak{L} = \partial_t + L; \quad L = b(x)\nabla + 2^{-1}\ a(x)\nabla^2 \tag{0.1}$$

where $a(x) = \sigma(x)\sigma(x)^T$ satisfies : $\mathfrak{L}^T = -\ \partial_t + L^T$ where

$$L^T = \delta(b.) + 2^{-1}\delta^2(a.) = -\ b'(x)\nabla + 2^{-1}a(x)\nabla^2 + c \tag{0.2}$$

with

$$b'(x) = -\ b(x) - \delta a(x)\ ;\ c(x) = \delta b(x) + 2^{-1}\delta^2\ a(x) \tag{0.3}$$

Assuming $\mathfrak{L}^T$ hypoelliptic, $M = m_t dt$ has a C^∞ density $m'(x,t)$ on int G ×]0,+∞[. For any fixed $t \geq 0$, m_t is the sum of the measure $m'_t = m'(x,t)dx$ on int G and of a simple sheet $q_t\delta_\Sigma$ supported by $\sum$. Hence the corresponding decomposition of M is

$$M = M' + Q\delta_\Sigma \tag{0.4}$$

Let $\sum_3$ be the non characteristic part of the boundary with respect to L^T and let

$$f(x) = b(x) + 2^{-1}\delta(a(x))$$

be the Fichera drift. Denoting by $\nu(x)$ the outward normal at $x \in \sum$, $\sum_2$, $\sum_0$ and $\sum_1$ are classically defined as the subsets of $\sum \setminus \sum_3$ where $f(x).\nu(x) < 0$, $= 0$ and > 0 resp. From the point of view of partial differential equations,the particular case $\sum = \sum_3$ is well known : then $Q = 0$, M' is solution of a Neumann problem and the smoothness of M' is known [O]. From the point of view of probability theory the case where $\sum' \cap \sum'' = \phi$ is clear ($\sum'$ and $\sum''$ are resp. the sticky part and the non attainable or immediately reflecting part of the boundary), but other cases are less clear. No example of reflected diffusion with $\mathfrak{L}^T$ hypoelliptic and discontinuous M' was known

(0.5) Example. Assuming the strong Markoff Property (SMP) and putting $\sum_{10} = \sum_1 \cup \sum_0$ then if

$$\textstyle\sum_3 = \phi, \qquad \bar{\bar{\sum}}_2 \cap \sum_{10} = \phi$$

let us prove that the partial differential equation determining M and Q are not really coupled.

In fact $\sum_{10}$ is sticky and $\sum_2$ is not attainable from int G a.s. Let τ be the first hitting time of $\sum_{10}$ by ξ with initial data $\xi(0) = x \in$ int G. Denoting by p_r the transition probability of the reflected process, SMP gives for any Borel subset B of G.

$$P(\xi(\tau+t) \in B||\mathcal{F}_\tau) = p_r(t,\xi(\tau),B) \tag{0.6}$$

Since $\sum_{10} = \sum_1 \cup \sum_0$ is sticky, p_r can be replaced in the RHS by the transition probability of the stochastic equation deduced from SDE by restricting it to $\sum_{10}$ and cancelling the normal drift. Hence denoting by Q' and Q" the parts of Q charging resp. $\sum_2$ and $\sum_{10}$, Q' is first characterized, then M is characterized in terms of Q', and finally Q" is characterized.

The purpose of the present lecture is to follow the study realized in [3] concerning the case where G is the semi-space $\{x_n \leq \beta\}$ with $\beta \geq 0$, where $\sum = \{x_3 = \beta)$ is characteristic ; and studying especially the case where $\bar{\sum}_2$ inversects $\sum_{10}$. Writing for $n \geq 2$

$$x = \begin{pmatrix} x' \\ x_n \end{pmatrix}, \quad \nabla = \begin{pmatrix} \nabla' \\ \partial_n \end{pmatrix}, \quad W = \begin{pmatrix} W' \\ W_n \end{pmatrix},$$

$$\xi = \begin{pmatrix} \xi' \\ \xi_n \end{pmatrix}, \quad b = \begin{pmatrix} b' \\ b_n \end{pmatrix}$$

we assume

$$\sigma(x) = \left(\begin{array}{c|c} \sigma'(x) & 0 \\ \hline 0 & 0 \end{array}\right) \tag{0.7}$$

Hence $\sum_2$, $\sum_0$ and $\sum_1$ are resp. the parts of boundary where $b_n(x',\beta) < 0$, $= 0$ and > 0. We only consider reflected diffusions ξ s.t.

$$\xi_0 \in \bigcap_{p>1} L^p(\Omega; \mathbb{R}^n) \tag{0.8}$$

$$m'_t \text{ (int G)} \to m'_0(\text{int G}) \quad \text{for } t \downarrow 0 \tag{0.9}$$

The last condition holds for ex if ξ is stationary, or if $\xi_0 = x \in$ int G. The following $\sum$-valued stochastic equation :

$$\text{SDE}' \quad d\xi'(t) = b(\xi'(t),\beta)dt + \sigma'(\xi'(t),\beta)dW'(t)$$

will be considered. The diffusion operator is $\mathfrak{L}' = \partial_t + L'$ with

$L' = b'(x',\beta)\nabla' + 2^{-1}a'(x)\nabla'^2$ and $a'(x) = \sigma'(x)\sigma'(x)$; the Fichera's drift is

$$f'(x) = b'(x',\beta) + 2^{-1}\delta'(a'(x',\beta)) \tag{0.10}$$

From [3], the diffusion operator of the reflected diffusion is $\mathfrak{L}_r = \partial_r + L_r$ with :

$$(L_r f)(x) = \begin{cases} L\ f(x) & \text{for } x \in G \setminus \sum_1 \\ L'\ f(x) & \text{for } x \in \sum_1 \end{cases} \tag{0.11}$$

The following generalized Fokker-Plank equation is established in [3]

GFDE $\quad \forall f \in \mathcal{D}(G \times \mathbb{R}) \quad \langle M, \mathfrak{L}_r f \rangle = - \langle m_o \delta_o, f \rangle$

We prove the equivalence of this equation with two coupled boundary value problems among which one at least is new

GFPE $\Leftrightarrow$ BVP1 + BVP2

The generalized FPE is also interpreted in terms of fluid-mechanics. This interpretation shows clearlywhy in the interesting cases BVP1 is not a usual first kind BVP [1] but a generalized problem of this type. The part IV gives a mathematical proof. Assuming in particular $|\sum_o| = 0$ and $\sum_3 \neq \phi$, we prove that :

a) the distribution-trace Θ of $b_n(x)m'(x,t)$ on $\sum \times \mathbb{R}$ cannot be a positive measure, hence a fortiori cannot be locally integrable.

b) $m'(x,t)$ cannot admit a distribution-trace on $\sum \times \mathbb{R}$. Hence a fortiori $m'(x,t)$ cannot be continuous on $G \times \mathbb{R}$.

The last part gives an æronautical application. A Comptes-Rendus Note of octobre 1984 gives a resumē of the present improved version of the oral lecture.

I. Preliminary probabilistic results.

These results are deduced borrowing the proofs of corresponding results for free SDE, using (0.11) and also the fact that a.s. the last component $\xi_n(t)$ admits for almost all t a right derivative of the following type :

$$D_+\xi_n(t) = \begin{cases} b_n(\xi(t)) \text{ if } \xi_n(t) \neq \beta \text{ or if } \xi_n(t) = \beta \text{ and } \xi'(t) \in \sum_{20} \\ \\ 0 \text{ if } \xi_n(t) = \beta \text{ and } \xi'(t) \in \sum_1 . \end{cases}$$

Hence $D_+\xi_n(t) \leq b_n(\xi(t))$ is all cases. Since $\beta \geq 0$

$$D_+(\xi_n(t)^{2m}) \leq \xi_n(t)^{2m-1} b_n(\xi(t)).$$

Hence in a standard way be combining Ito's formula and Gronwall's Lemma we obtain :

(I.2) $\forall m$ integer ≥ 1 $\forall T > 0$: $\sup_{0\leq t\leq T} E|\xi(t)|^{2m} < \infty$

(I.3). Theorem. *Suppose* σ *is constant and* $b(x)$ *linear. Then :*

(a) *If* $b(x).x \leq -c\|x\|^2$ *for all* $x \in \mathbb{R}^n$ *and some* $c < 0$, *there exists a stationary solution of* SDE *reflected in* $G = \{x_n \leq \beta\}$.

(b) *If* SDE *is reflected by the boundary of the strip* $\{\alpha \leq x_n \leq \beta\}$, *there exists a stationary solution if* $b_1' x'-x' \leq -c'\|x\|^2$ *for all* $x' \in \mathbb{R}^{n-1}$ *and some* $c' < 0$ *; where* b_1' *is deduced from* b *by suppressing the last column and the last row* .

(I.4) Lemma.

(a) *The Markoff process defined by the reflection of* SDE *on the boundary of a closed convex subset* G *of* $\mathbb{R}^n$ *satisfies for* T *finite* > 0 *and for arbitrary* x *and* $x' \in G$

(I.5) $\sup_{0\le t\le T} E|\xi_x(t)-\xi_{x'}(t)|^2 \le C|x-x'|^2$

where C *depends only on* G,T *and on the coefficients of* SDE.

(b) *The Feller's property is satisfied and*

(I.6) $\varepsilon > 0 \quad p(t,x,B(x,\varepsilon)) \to 1$ *if* $t \downarrow 0$

Proof. a) A deduction similar to [3] p. 588 by using monotonicity and Ito's formula for $0 \le t \le T$ gives

(I.7) $E|\xi_x(t)-\xi_{x'}(t)|^2 \le C|x-x'|^2 + C\int_0^t E|\xi_x(s)\xi_{x'}(s)|^2\,ds$

with C as in the lemma. Then (I.5) follows using Gronwall's lemma.

b) Let $t > 0$ and let f be a real-valued bounded continuous fonction on G. By a) the mapping $x \to \xi_x(t)$ in continuous $G \to L^2(\Omega,\mathbb{R}^n)$. Since the L^2-convergence of random variables implies the narrow-convergence of laws, the following mapping is continuous

$$G \ni x \to (T_t f)(x) = E[f(\xi_x(t))].$$

This proves Feller's property. Finally (1.6) results from the narrow-continuity of $t \to \text{Law}\ \xi_x(t)$.

(I.8) Lemma. *Under hypothesis of theorem* (1.3)

$$T^{-1}\int_0^T p(t,x,B(R)^c)dt \to 0 \text{ if } R \to \infty$$

uniformly in $T > 0$, *where* B(R) *denotes the intersection of* G *with the ball* $\{\|x\| \le R\}$.

Proof. We first prove the existance of a C^2-function V on G s.t. $V \ge 0$ and

$$A_R = \sup_{|x|\ge R} L_r V(x) \to -\infty, \quad \text{if } R \to \infty.$$

Take for instance $V(x) = \|x\|^2$ in the case a) and $V(x) = \|x'\|^2$ in case b).Then lemma (1.8) follows from a) and (0.11) borrowing the argument from [2] p. 90.

In view of previous results the theorem (I.3) follows from [2] theorem 2.1 p. 71

II. TRACE THEOREM AND SPLITTING OF THE GENERALIZED FPE

Let $\dot{\mathcal{B}} = \dot{\mathcal{B}}(\mathbb{R}^n)$ be the space of all C^∞-function $f(x)$ on $\mathbb{R}^n$ s.t. for any multi-index $\alpha = (\alpha_1,\ldots,\alpha_n)$: $\partial^\alpha f(x) \to 0$ for $|x| \to \infty$. This is a Frechet space for the semi-norms $\|f\|_\infty = \sup_x |\partial^\alpha f(x)|$. Hence $O_c = O_c(\mathbb{R}^n) = \cup_{k=1}^\infty (1+|x|^2)^k \dot{\mathcal{B}}$ is like $\mathcal{D}$ an inductive limite of Frechet spaces.

The space $OC(\mathbb{R}^{n+1}) = OC$ of test functions. For any relatif integer $B_k = B_k(\mathbb{R}^{n+1})$ denotes the space of all C^∞-function $f(x,t)$ on $\mathbb{R}^{n+1}$ vanishing for $t \geq k$ such that for any pair (α,β), $\partial^{\alpha,\beta} f = \partial_x^\alpha \partial_t^\beta f(x,t) \to 0$ for $|x| \to \infty$ uniformly for t in any compact intervall K . This is a Frechet space for the semi-norms $\|f\|_{\alpha,\beta,K} = \sup_{x,t\in K} |\partial^{\alpha,\beta} f(x,t)|$. Hence the space

$$OC_k = \cup_{\ell=1}^\infty (1+|x|^2)^\ell \, B_k$$

is like O_c an inductive limit of Frechet spaces. Finally $OC = \bigcup OC_k$ is the inductive limit of the spaces OC_k ; OC is a normal space of distributions i.e. the canonical injections :

$$\mathcal{D}(\mathbb{R}^{n+1}) \subset OC \subset \mathcal{D}'(\mathbb{R}^{n+1})$$

are continuous with dense range. The corresponding retrograd space OC_R is defined as the space of all function $f(x,-t)$, $f \in OC$.

(II.1) The space OC' of distributions on $\mathbb{R}^{n+1}$ is defined as the dual of OC. Note that any $u \in OC'$ vanishes on some half space $t \leq t_0$ i.e. for t sufficiently negative. The space $OC'(\sum \times \mathbb{R})$ is defined in the same way. These spaces can be used here since by (I.2) the extension $\widetilde{M}$ of M by zero outside $G \times \mathbb{R}$ belongs to $OC'(\mathbb{R}^{n+1})$. In the same way $Q \in OC'(\sum \times \mathbb{R})$. Since $\mathfrak{L}$ is continuous on $OC(\mathbb{R}^{n+1})$, $\mathfrak{L}^T$ defines a weakly continuous mapping on $OC'(\mathbb{R}^{n+1})$.

For an arbitrary real w, $k_w(t)$ denotes the cutoff function on $\mathbb{R}$ vanishing for $t \leq w$ and $k_w(t) = 1$ for $t \geq w$. The following result is a variant of [4] propositions (27) and (29).

(II.2) <u>Trace theorem for</u> $b_n M'$. The measures

$$\Theta_{vw}(dx',dt) = b_n(x',v)k_w(t)m'(x',v,t)dx'dt$$

tend for $v\uparrow\beta$ and $w\downarrow 0$ in the weak space $OC'(\sum \times \mathbb{R})$ to some limit Θ. Denoting $\tilde{m}'_o$ the extension of m'_o by 0 outside G, the following equality holds in $OC'(\mathbb{R}^{n+1})$

$$\text{GFPE}' \quad \mathfrak{L}^T(\tilde{M}') = -\tilde{m}'_o(dx)\delta_o(t) + \Theta(x',t)\delta_\beta(x_n)$$

<u>Proof</u>. Let M'_{vw} be the restriction of $M' = m'(x,t)dxdt$ to $U(v,w) = \{(x,t) \in \mathbb{R}^n, x_n \leq v \text{ and } t \geq w\}$ and let $\tilde{M}_{vw}$ be the extension of M'_{vw} by zero outside U(v,w). The Green's formula applied to M', to $\mathfrak{L}$ and to an arbitrary test function $\in \mathfrak{D}(\mathbb{R}^{n+1})$ gives the following equality in $\mathfrak{D}'(\mathbb{R}^{n+1})$

$$\mathfrak{L}^T(\tilde{M}'_{vw}) + (\tilde{m}'_w k_v(x_n))\delta_w(t) = \Theta_{vw}(x',t)\delta_v(x_n).$$

Since the LHS belongs to $OC'(\mathbb{R}^{n+1})$, this is an equality in $OC'(\mathbb{R}^{n+1})$. For $v \uparrow \beta$ and $w \downarrow 0$, the LHS tends weakly to $\mathfrak{L}^T(\tilde{M}') + \tilde{m}'_o\delta_o(t)$ in view of (I.2). Hence the RHS tends weakly to the same limit T. The application of this result to x_n-independent test functions shows that Θ_{vw} tends weakly in $OC'(\sum \times \mathbb{R})$ to some limit Θ s.t. $T = \Theta(x',t)\delta_\beta(x_n)$. Hence the trace- theorem is proven. Bellow Θ is called the distribution- trace of $b_n M'$ on $\sum \times \mathbb{R}$.

(II.3) <u>Splitting theorem for Generalized FPE</u>. The Generalized FPE is equivalent with BVP1 + BVP2 where

<u>BVP1</u> : the C^∞-function $m'(x,t)$ satisfies

(II.4) $\quad \mathfrak{L}^T M' = 0$ on int $G \times]0,+\infty[$

and the distribution trace Θ of $b_n M'$ satisfies

(II.5) $\Theta = q_0(x')\delta_0(t_n)$ on $\sum_2 \times \mathbb{R}$

(II.6) Moreover Θ is positive on $\sum_1 \times \mathbb{R}$

BVP2 : The positive measure Q on $\sum \times \mathbb{R}$ vanishes for $t < 0$ and on $\sum_2 \times]0,+\infty[$. Moreover the following equality holds in $OC'(\sum \times \mathbb{R})$

GFPE" $\quad (\mathfrak{L}'^T Q)(x',t) = - \;\Theta(x,t) - q_0(x')\delta_0(t).$

Proof. Since $M = M' + Q\delta_\Sigma$ and $m_0 = m_0' + q_0\delta_0$ GFPE can be written :

$$-\int \phi(x,0)(m_0'(dx) + q_0(dx'))\delta_\beta(x_n) \;=\; \langle M',\mathfrak{L}_r\phi\rangle + \langle Q\delta_\beta(x_n),\mathfrak{L}_r\phi\rangle.$$

Since $\mathfrak{L}_r = \mathfrak{L}$ on int $G \times \mathbb{R}$, the first term of the RHS can be evaluated using GFPE, hence after cancellation :

$$-\int \phi(x',\beta,0)q_0(dx') = \langle\Theta(x',t)\delta_\beta(x_n),\phi\rangle + \langle Q\delta_\beta(x_n),\, \mathfrak{L}_r\phi\rangle$$

In this identity, a normal derivative $b_n(x)\partial\phi/\partial x_n$ of ϕ appears in the last term on $\sum_2 \times \mathbb{R}$. Hence the corresponding term must vanish i.e. Q vanishes on $\sum_2 \times \mathbb{R}$. Now last identity depends only on the trace $\psi(x',t) = \phi(x',\beta,t)$ of ϕ on $\sum \times \mathbb{R}$. Hence the last identity can be written :

$$-\int \psi(x',0)q_0(dx') = \;\langle\Theta(x',t),\psi\rangle + \langle Q,\, \mathfrak{L}'\psi\rangle$$

for an arbitrary $\psi \times \mathcal{D}(\sum \times \mathbb{R})$ and this proves (GFPE)'. Since Q vanishes on $\sum_2 \times \mathbb{R}$ and since the differential operator $\mathfrak{L}'^T$ does not increase supports (GFPE)" shows that Θ is supported by $\sum_{10} \times [0,+\infty[$. Finally Θ is a positive measure on $\sum_1 \times \mathbb{R}$ since the distributions- limit of a sequence of positive measures is a positive measure.

(II.7) The results of this part have a simplified form for stationary processes since then $\Theta(x',t) = \Theta(x')dt$, $M' = m'(x)dt$... and since the time variable disappears. The Generalized FPE is now equivalent with the following condition for $m(x) = m'(x) + q(x')\delta_\beta(x_n)$

bvp1 : The measures $m'(x',v)b_n(x',v')dx'$ converge weakly in $\mathcal{O}_c'(\mathbb{R}^{n-1})$ to some limit θ for $v \uparrow \beta$ and

$L^T m' = 0$ in int G ; $\theta = 0$ on $\sum_2$

bvp2 : The positive measure q on $\sum$ vaniskes on $\sum_2$ and satisfies

$L'^T q = \theta$ in $\mathcal{O}_c'(\mathbb{R}^{n-1})$.

Extending the terminology of [5] to second order operators with positive symbol [6], BVP1 and bvp1 appear as "generalized" first kind BVP.

III. GENERALIZED FPE AND FLUID MECHANICS.

As known by Planck, Schrodinger..., elimination of the momentum density VM' between :

(III.1) $\partial_t M' - \delta(VM') = 0$ and $VM' = fM' - 2^{-1}a(\nabla M')$

gives the Fokker-Planck Equation $\mathcal{L}^T M' = 0$ Equation (III.1) are resp. a continuity equation and a law giving the momentum density VM in term of the mass density M'. Hence for any closed regular domain B of in G, the global transfer rate of mass through the hypersurface ∂B is $\int_{\partial B} V(x)M'(x).\vec{n}(x)d\sigma(x)$.
In the same way EFPG' can be obtained for $t > 0$ by eliminating the momentum density V'Q betwenn

(III.2) $\partial_t Q - \delta'(V'Q) = \Theta$ and $V'Q = f'Q - 2^{-1}a'(\nabla'Q)$

Hence by the Equivalence Theorem, Generalized FPE represents the motion of a fluid moving in the following way in G for $t \geq 0$. For initial time $t = 0$, the mass distribution is $m = m_o + q_o\delta_\beta(x_n)$. For $t > 0$ the fluid is moving in int G according to (III.1). Since int G is the limit for $v \uparrow \beta$ of the half spaces $x_n \leq v$, the local rate of mass transfer from int G to $\sum$ is characterized by Θ. The positivity of Θ on $\sum_1 \times]0,+\infty[$ means that masses are moving for $t > 0$ from int G to $\sum_1$. The vanishing of Θ on $\sum_2 \times]0,+\infty[$ means that no transfer of masses arises through $\sum_2$ for $t > 0$. The fluid is moving in $\sum$ according to (III.2). These two equations describe the mass diffusion defined by EDS' and the phenomenon of mass exchange with int G. The total mass 1 is preserved during the motion since Θ appears in (III.2) as a source term.

(III.3) <u>In the stationary case</u>, there is no global mass exchange between int G and $\sum$, hence

(III.4) $\int_\Sigma \theta(x)dx = \langle\theta,1\rangle = 0$

This means that all masses getting up from int G to $\sum_1$ and then moving in $\sum_{10}$ are coming back in int G along $\sum_0$.

(III.5) <u>Fact</u> : Suppose $\theta \neq 0$. Then m' cannot have a distribution trace on $\sum$. If moreover $|\sum_0| = \int_{\Sigma_0} dx' = 0$, then θ cannot be locally integrable. In fact suppose that $m'(x',v)dx'$ tends weakly to some distribution ℓ for $v \uparrow \beta$. Since $m'(x',v)dx'$ is a positive measure, ℓ is necessarily positive. Then $b_n(x',v)m'(x',v)dx'$ tends to $\theta = b_n(x',\beta)\ell(x')$. Hence θ vanishes on $\sum_{02}$. Hence $\theta \geq 0$ and therefore $\theta = 0$ by (III.4) : which is impossible. If moreover $|\sum_0| = 0$, θ cannot be locally integrable since then $\theta \geq 0$, hence $\theta = 0$ by (III.4) : this is impossible.

The fact (III.5) means that we have a kind of "waterfall phenomenon" if G : the probabilistic fluid is identified with water and $\sum_0$ represents the part of the horizontal surface $\sum_{10}$ where water falls in int G. If the size of $\sum_0$ is too small, m' and $b_n m'$ cannot be smooth. Let us prove mathematically these results :

IV. IRREGULARITY OF THE SOLUTION OF GENERALIZED FPE.

In order to find a left inverse of $\mathfrak{L}^T = -\partial_+ + L^T$ the following auxiliary stochastic equation is considered

$$\text{SDE"} \qquad d\zeta(t) = b'(\zeta(t))dt + \sigma(\zeta(t)).dW(t)$$

where b' is defined by (0.3). The diffusion operator of this equation is $\mathcal{M} = \partial_t + M$ with

$$\text{(IV.I)} \qquad M = b'(x).\nabla + 2^{-1}a(x).\nabla^2$$

For arbitrary $x \in \mathbb{R}^n$, ζ_x denotes the solution of SDE" starting at time zero from x. For an arbitrary $(x,t) \in \mathbb{R}^{n+1}$, $T_{x,t} = T_{x,t}(y,u)$ denotes the positive measure on $\mathbb{R}^{n+1}$ vanishing for $u < t$ and such that for any $\phi \in OC$

$$\text{(IV.2)} \quad \langle T_{x,t},\phi\rangle = E[\int_t^\infty du\phi(\zeta_x(u-t,u)e^{\int_0^{u-t} c(\zeta_x(s))ds}]$$

where c is defined by (0.3).

(IV.3) Lemma. The kernel $T = \{T_{x,t},(x,t) \in \mathbb{R}^{n+1}\}$ is a left inverse of $-(\mathcal{M}+c)$ and more precisely

$$\text{(IV.4)} \qquad \phi \in OC(\mathbb{R}^{n+1}) \quad -\phi(x,t) = T_{x,t}(\mathcal{M}\phi+c\phi)$$

Principle of the proof. The following $\mathbb{R}^{n+1}$ - valued stochastic equation in $\underline{\zeta} = (\zeta,\zeta_n)$:

$$d\zeta(t) = b'(\zeta(t))dt + \sigma(\zeta(t))dW(t)$$

$$d\zeta_{n+1}(t) = c(\zeta(t))\zeta_{n+1}(t)dt$$

have the diffusion operator $\mathcal{M}' = \mathcal{M} + c(x)x_{n+1}\partial_{n+1}$. The solution $\underline{\zeta}_x$ of this stochastic equation starting at time zero from $\underline{x} = (x,x_{n+1})$ has for its last component

$$\zeta_{n+1}(t) = x_{n+1} . \exp \int_0^t c(\zeta_x(u))du$$

Putting $\psi(\underline{x},t) = \phi(x,t)x_n$, an application of Ito's formula to $u \to \psi(\underline{\zeta_x}(u-t),u)$ gives

$$- \psi(\underline{x},t) = E\left[\int_0^\infty du(\mathcal{M}'\psi)(\underline{\zeta_x}(u-t),u)\right]$$

$$= x_n \; E\left[\int_t^\infty du \; (\mathcal{M}\phi)(\zeta_x(u-t),u)e^{\cdots} + c(\zeta_x(u-t))e^{\cdots}\}\right]$$

hence gives (IV.4) for $x_n = 1$. Since the substitution $(x,t) \to (x,-t)$ transforms $\mathcal{M} + c$ in $\mathfrak{L}^T$, the previous lemma implies the following result

(IV.5) Lemma. *For an arbitrary* $(x,t) \in \mathbb{R}^{n+1}$, $S_{xt} = S_{x,t}(y,u)$ *denotes the positive measure on* $\mathbb{R}^{n+1}$ *vanishing for* $u > t$ *s.t.* $\phi \in OC_R(\mathbb{R}^{n+1})$

$$\text{(IV.6)} \quad \langle S_{x,t},\phi\rangle = E\left[\int_{-\infty}^t du \; \phi(\zeta_x(t-u),u)e^{\int_0^{t-u} c(\zeta_x(s))ds}\right]$$

Then the kernel $S = (S_{xt})$ *is a left inverse of* $- \mathfrak{L}^T$ *in* $OC_R(\mathbb{R}^{n+1})$

$$\text{(IV.7)} \qquad \phi \in OC_R(\mathbb{R}^{n+1}) \qquad -\phi(x,t) = S_{x,t}(\mathfrak{L}^T\phi).$$

In the same way $- \mathfrak{L}'^T$ admits a left inverse $S' = (S'_{x',t})$ in $OC_R(\sum \times \mathbb{R})$.

(IV.8) The irregularity theorem. Let $\xi = \{\xi_t, \; t \geq 0\}$ be a solution of SDE reflected by the boundary $\sum = \{x_n = \beta\}$ of the half space G of $\mathbb{R}^n$ where $x_n \leq \beta$. We assume (0.7), (0.8), (0.9) and also a) + b) + c) where :

a) $|\sum_0| = 0$ and $\sum_2 \neq \emptyset$

b) The left inverse S and S' of $\mathfrak{L}^T$ and $\mathfrak{L}'^T$ defined by (IV.5) admit weakly continuous extensions in $OC'(\mathbb{R}^{n+1})$ and $OC'(\mathbb{R}^n)$ resp.

c) For arbitrary t, $u \leq t$, $x \in \mathbb{R}^n$ and $x' \in \mathbb{R}^{n-1}$ the measures $S_{x,t}(dy,u)$

and $S'_{x',t}(dy',u)$ have C^∞ and strictly positive densities.

Then the distribution-trace Θ of $b_n m'$ on $\sum \times \mathbb{R}$ defined by the trace theorem cannot be a measure ≥ 0. Hence in particular : $\Theta \neq 0$, Θ cannot be locally integrable and m' does not admit a distribution - trace on $\sum \times \mathbb{R}$.

In particular m' cannot be continuous on $G \times]0,+\infty[$.

Proof of the theorem. Let us assume $\Theta \geq 0$. In view of the hypothesis b) the equation GFPE' and GFPE" can be inverted using S and S' resp. :

This gives

$$\text{(IV.9)} \qquad \tilde{M}' = - S(\Theta\delta_\beta + \tilde{m}_o\delta_o)$$

$$\text{(IV.10)} \qquad Q = - S'(\Theta + q_o)$$

If $\Theta + q_o \neq 0$, the last relation gives for an arbitrary relatively compact open subset w' of $\sum_2$ and t large enough.

$$\inf_{x' \in w'} Q(x',t) > 0 \text{ .}$$

This is impossible since Q vanishes on $\sum_2 \times]0,+\infty[$. Hence necessar ly $\Theta + q_o = 0$. Since we assume $\Theta \geq 0$, the last equality implies $\Theta = q_o = 0$ i.e. $m'_o \neq 0$. Therefore (IV.9) gives for an arbitrarely relatively compact subset w of $G^c = \mathbb{R}^n \setminus G$ and $t > 0$:

$$\inf_{x \in w} \tilde{M}'(x,t) > 0$$

This is impossible since $\tilde{M}'$ vanishes outside $G \times]0,+\infty[$. In conclusion Θ cannot be a measure ≥ 0.

If $\Theta \geq 0$ is impossible, $\Theta = 0$ is also impossible. Since $|\sum|_o = 0$ and since $\Theta \geq 0$ on $\sum_1 \times \mathbb{R}$ the hypothesis " Θ locally integrable" implies $\Theta \geq 0$: this is impossible. Finally suppose m' admits a distribution trace d on $\sum \times \mathbb{R}$. Since $m' \geq 0$, d is necessarely ≥ 0. Since b_n vanishes on $\sum_o$, the trace Θ of $b_n m'$ on $\sum \times \mathbb{R}$ would be zero on $\sum_{20} \times \mathbb{R}$. Hence $\Theta \geq 0$: this is impossible.

(IV.11) Theorem. *The hypothesis a), b) and c) of the irregularity theorem are satisfied if the following conditions are satisfied*

(i) σ *constant*

(ii) $b(x) = bx + b_0$ *is a linear function of* x

(iii) *The subset of* $\sum$ *where* $b_n(x', \beta) = 0$ *is an hyperplane of* $\sum$

(iv) *The following two matrices of order* $n \times n^2$ *and* $(n-1) \times (n-1)^2$ *resp.*

$$M = [a, ab, \ldots ab^{n-1}] \quad \textit{and} \quad M' = [a', a'b_1' \ldots a'b_1'^{n-2}]$$

have rank n *and* $(n-1)$ *respectively.*

Proof. Since $b'(x) = -b(x) = -bx - b_0$ and $c = -\mathrm{Tr}\, b$, the solution ζ_x of the auxiliary stochastic equation $d\zeta = -b(\zeta)dt + \sigma\, dW$ is for $u > t$.

$$\text{(IV.12)} \qquad \zeta_x(u) = e^{-bu}x + \int_0^u e^{-b(u-s)}(\sigma\, dW(s) - b_0 ds)$$

Hence Law $\zeta_x(u) = \mathrm{Gauss}(\Gamma_u, m_u)$ is Gaussian with mean

$$\text{(IV.13)} \qquad m_u = (\exp -bu)x - \int_0^u (\exp -b(u-s))b_0 ds \text{ and covariance}$$

$$\text{(IV.14)} \qquad \Gamma_u = \int_0^u e^{-b(u-s)} \sigma\sigma^t e^{-b^T(u-s)} ds$$

In our particular case, (IV.2) is reduced in :

$$\text{(IV.15)} \qquad \langle T_{x,t}, \phi \rangle = E\left[\int_t^\infty e^{-(u-t)c} \phi(e^{-bu}x + \int_0^u \ldots) du\right]$$

Hence for $u > t$

$$\text{(IV.16)} \qquad T(x,t,.,u) = e^{-c(u-t)} \mathrm{Gauss}(\Gamma_u, m_u)$$

As well known, Γ_u is invertible for all $u > 0$ iff Rank $M = n$. Hence in our case the kernel of S (or of T) is C^∞ and > 0 outside the diagonal. In the same way the hypothesis Rank $M' = n-1$ implies that the kernel of S' is C^∞ and > 0 outside the diagonal of $\mathbb{R}^n \times \mathbb{R}^n$. The previous results are implicit in [2]'. In view of (i) the condition a) of the irregularity theorem is satisfied. Hence we only have to check that the transpose T^T of the kernel T defines a linear and continuous mapping in OC_R. Note that the procedure described before in order to find a left fundamental solution of the diffusion operator $\mathcal{M}$ of SDE" can be applied to the stochastic equation $d\xi = b\xi + \sigma\, dW$. Hence a left fundamental solution U of $\mathcal{L} = \partial_t + bx.\nabla + 2^{-1}a.\nabla^2$ is constructed s.t. for any $f \in OC$, $\tilde{f} = U f$ is explicitely given by

$$\text{(IV.17)} \qquad \tilde{f}(x,t) = \int_t^\infty du\; E[f(\xi_{x,t}(u),u)]$$

where ξ_{xt} starts from w at time t. Hence in terms of the process ξ_x starting from x at time zero :

$$\tilde{f}(x,t) = \int_0^\infty du\; E[f(\xi_x(u),t+u)]$$

$$\text{(IV.18)} \qquad \tilde{f}(x,t) = \int_0^\infty du\; E[f(e^{bu}x + \int_0^u e^{(u-s)b}\sigma\, dW(s),t+u)]$$

A straitforward computation shows : a) T is also a right inverse of $-(\mathcal{M}+c)$

$$-(\mathcal{M}+c)T = Id \Longleftrightarrow -T^T(\mathcal{M}+c)^T = Id$$

b) Moreover T^T is constructed by the probabilistic technique described previously. Hence finally the theorem follows from the following result :

(IV.19) <u>Lemma</u>. *The mapping* $U : f \to \tilde{f}$ *defined by* (II.18) *is linear and continuous in* OC.

Since U maps OC_k in OC_k for arbitrary k, we only cheek for any ℓ and for convenient ℓ', U defines a continuous mapping

(IV.20) $\omega^{\ell} B_k \to \omega^{\ell'} B_k$ with $\omega(x) = (1+|x|^2)$

Putting $\partial^{\alpha,\beta} \tilde{f} = \partial_x^{\alpha} \partial_t^{\beta} \tilde{f}$ we have

$$\partial^{\alpha,\beta} \tilde{f}(x,t) = \int_0^k du \; E[\sum_{|\gamma|=\beta} a_\gamma(u) \partial^{\gamma,\beta} f(e^{bu}x + \int_0^u \ldots, t+u)]$$

where $a_\gamma(u)$ denotes a polynomial function of the coefficient of exp bu with d$^{\underline{o}}$ $a_\gamma(u) \le \beta$. Since f tends to the origin of $\omega^{\ell} B_k$:

$$\partial^{\gamma,\beta} f(x,s) = \omega(x)^{\ell} g_{\gamma,\beta}(x,s)$$

where $g_{\gamma,\beta}$ tends to the origin of B_k. Hence

$$|\partial^{\alpha,\beta} f(x,t)| \le \sum_{|\gamma|=\beta} \|g_{\gamma,\beta}\|_\infty \int_0^{k+t} |a_\gamma(u)| \; E[\omega^{\ell}(e^{bu}x + \int_0^u \ldots)]du$$

$$\le \sum_{|\gamma|=\beta} \|g_{\gamma,\beta}\|_\infty \int_0^{k+t} |a_\gamma(u)| du \int_{\mathbb{R}^n} (1+|e^{bu}x + y|^2) \; g_u(dy)$$

where g_u denotes a centered Gaussian measure on $\mathbb{R}^n$ with covariance

$$\int_0^u e^{(u-s)b} \sigma\sigma^T e^{(u-s)b^T} ds$$

Hence the continuity of (IV.20) follows for $\ell' = \ell$ using the inequality

$$(1+|e^{bu}x+y|^2)^{\ell} \le c(1+e^{bu}|x|^2)^{\ell} (1+|y|^2)^{\ell}$$

This proves the lemma hence also the theorem (IV.11). This theorem gives a large class of reflected diffusions for which the irregularity theorem shows also the existence and uniqueness of $u \in OC'$ solution of

$$\mathcal{L}^T u = f \quad ; \; f \text{ given} \in OC'$$

for $\mathcal{L} = \partial_t + (bx+b_0).\nabla + \sigma.\sigma^T.\nabla^2$ if Rank M = n.

(IV.20) Remark. In the irregularity theorem, the hypothesis concerning $\mathcal{L}^T$ and $\mathcal{L}'^T$ are stronger that the hypoellipticity : we also assume the existence of left fundamental solutions continuous in OC' and strictly positive outside the diagonal. For example σ constant, b linear, Rank M = n, rank M' = n-1, the hypoellipticity of $\mathcal{L}^T$ and $\mathcal{L}'^T$ follows directly from [1] but other properties have to be proven.

V. AN AERONAUTICAL APPLICATION.

Francis Poirion of ONERA has studied in [7] a simplified model for the automatic control of airfoils of an airplane flying in the turbulent wind. The vertical turbulency is modelized by a stationary Ornstein-Uhlenbeck process (z_t) solution of

$$(V.1) \qquad dz_t = -\alpha \; z_t dt + \beta \; dW_t \quad ; \qquad \alpha \text{ and } \beta > 0$$

Since the motion of an airfoil has (for example) a limited amplitude, this motion is modelized by a scalar process (y_t) such that $|y_t| \leq Y_o$ for all t and such that

$$(V.2) \qquad dy_t = -(\gamma y_t + \delta z_t)dt \qquad \text{for} \quad |y_t| < Y_o$$

Substituing z_t by $x_t = \gamma y_t + \delta z_t$, Francis Poirion obtains a diffusion $\xi_t = \begin{pmatrix} x_t \\ y_t \end{pmatrix}$ living in the strip $G = \{|y| \leq Y_o\}$ solution in int G of the system

$$(V.3) \qquad \begin{cases} dx_t = -(ax_t + by_t)dt + c\; dW_t \\ dy_t = x_t dt \qquad ; \text{ with a,b and } c > 0 \end{cases}$$

and reflected by the boundary $\sum = \sum^+ \cup \sum^-$ with $\sum^{\pm} = \{y = \pm Y_o\}$. By theorem (I.3) there exists a stationary solution (ξ_t). But all results of part II, III, IV established for an half space $G = \{x_n \leq \beta\}$ are also valid if G is a strip $\{\beta' \leq x_n \leq \beta\}$. In particular (2.7) shows that the probability law $m' + q\delta_\Sigma$ of ξ_t is solution of a system bvp1 + bvp2 of coupled partial differential equations. Since the conditions (i) ... (iv) of (IV.11) are

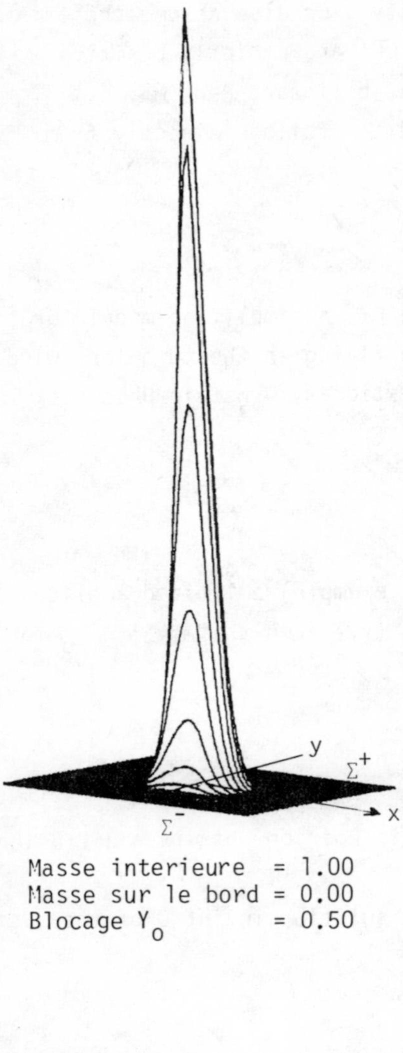

Masse interieure = 1.00
Masse sur le bord = 0.00
Blocage Y_o = 0.50

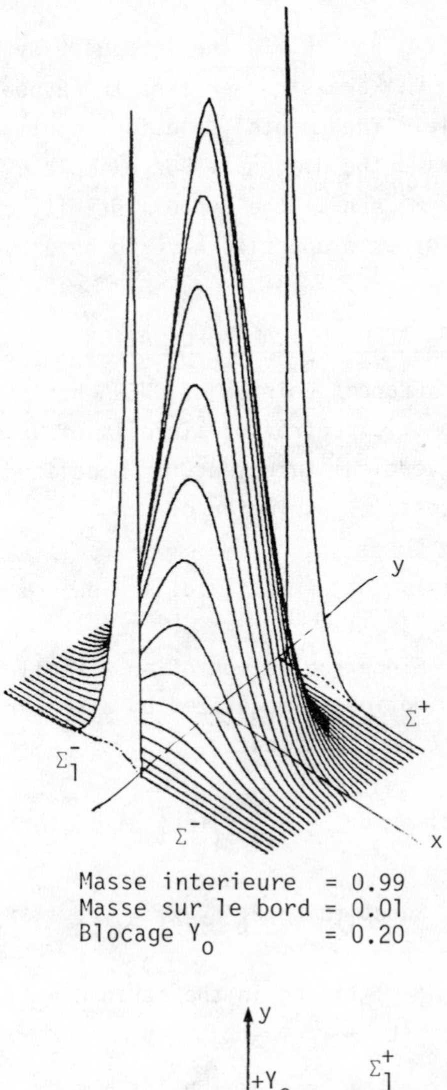

Masse interieure = 0.99
Masse sur le bord = 0.01
Blocage Y_o = 0.20

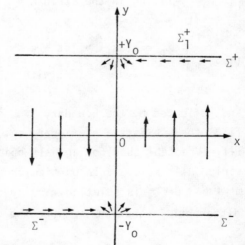

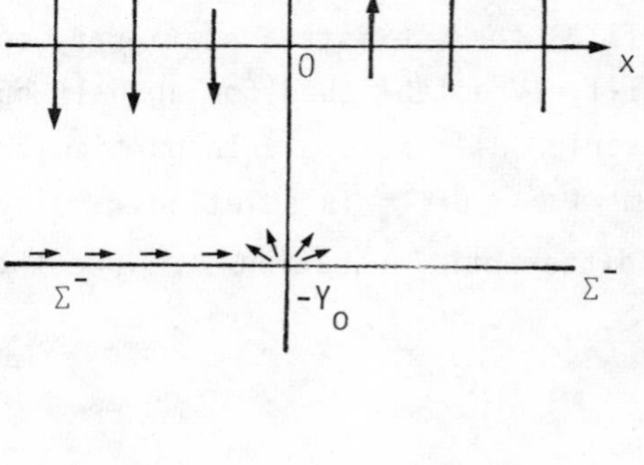

Figure 1 Computation of the probability law of a stationary reflected two-dimensional diffusion by F. POIRION (O.N.E.R.A.)

The diffusion $\begin{cases} dx = -(ax + by)\,dt + dw \\ dy = x\,dt \end{cases}$

is reflected by the boundary

$\Sigma = \Sigma^+ \cup \Sigma^- = \{y = \pm Y_o\}$

of the horizontal strip G where $|y| \le Y_o$.

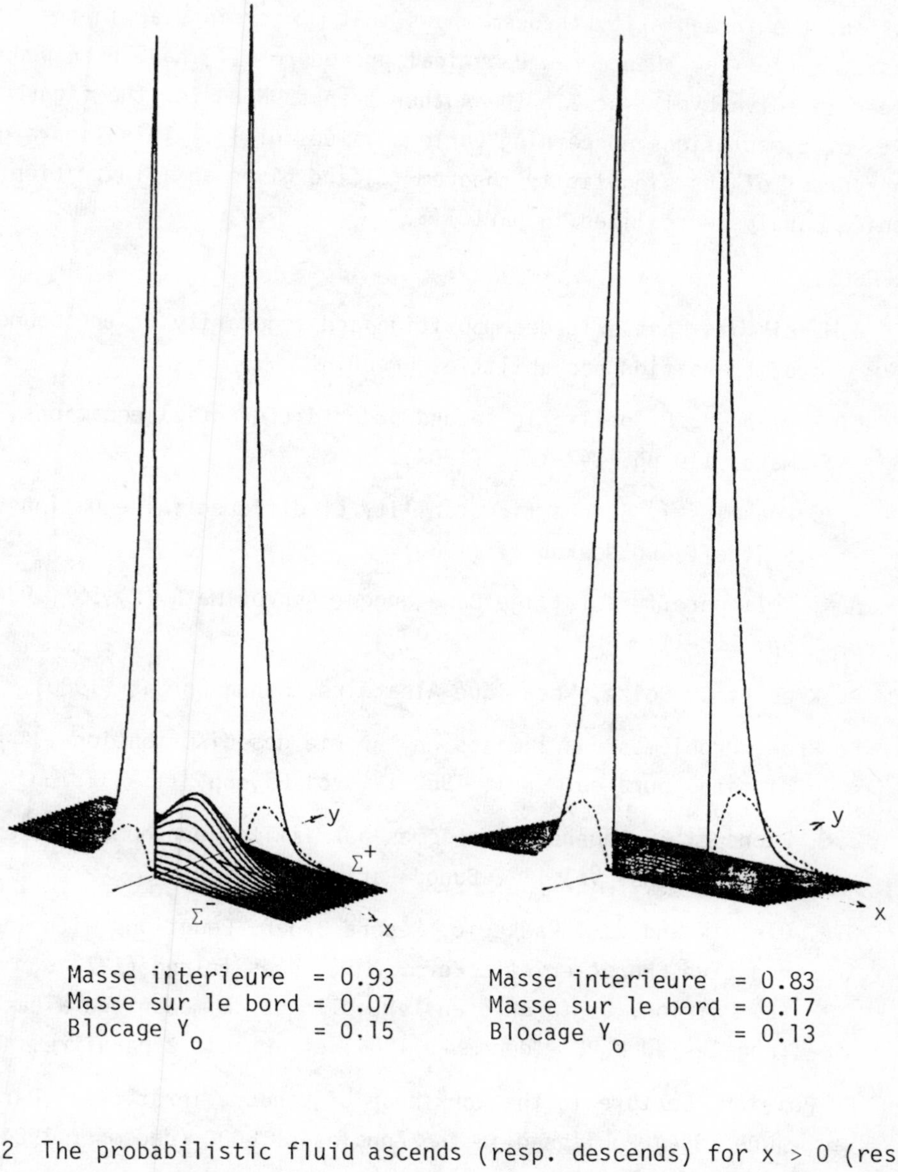

Masse interieure	= 0.93	Masse interieure	= 0.83
Masse sur le bord	= 0.07	Masse sur le bord	= 0.17
Blocage Y_o	= 0.15	Blocage Y_o	= 0.13

Figure 2 The probabilistic fluid ascends (resp. descends) for $x > 0$ (resp. $x < 0$), then goes to the left at Σ^+_- (resp. to the right at Σ^-_1) and falls to int G at the two points $(0, \pm Y_o)$. We have $m = m' + q\,\delta_\Sigma$ where m' (resp. $q\,\delta_\Sigma$) is the part of m concentrated on int G (resp. on Σ). The four images give the graphs of m' (and also of q in dotted lines) for various values of Y_o. In the left-hand picture Y_o is large, hence $q \simeq 0$ and m' is $\simeq$ gaussian. The illustrations show how the singularity of m' increases when Y_o decreases.

satisfied, the irregularity theorem shows that m' is singular in the two points $\Sigma_0^{\pm} = (0, \pm Y_0)$. Hence new numerical procedures [7] have been proposed in order to solve bvpl + bcp2. The Author thanks ONERA for the figure giving results of computations concerning various values of Y_0. This figure gives another proof of the singularity theorem (!) and gives an illustration of the mechanical analysis realized in part III.

REFERENCES.

[0] J.M. Bismut, Last exit decomposition and regularity at the boundary of transition probabilities Preprint.

[1] L. Hörmander, Hypoelliptic second order differential equations. Acta Math. 119 pp. 147-171, (1967).

[2] R.Z. Khasminski, Stochastic stability of differential equations. Sijthoff and Noordhoff (1980).

[2'] A.N. Kolmogoroff, Zufällige Bewegungen. Ann of Math (2) 35 (1934) pp. 116-117.

[3] P. Kree et C. Soize, Mecanique Aleatoire, Dunod Bordas (1980).

[4] P. Kree, Problêmes aux limites en théorie des distributions. Annali di math. pura ed. appl. 1969 IV. vol 83, pp. 113-132.

[5] J.L. Lions et E. Magenes, Problèmes aux limites non homogênes et applications. Vol 1+2. Dunod Paris 1968-1969.

[6] O.A. Oleinik and E.V. Radkevic. Second Order. Equations with Non-negative Characteristic Form. AMS. Rhods Island (1973).
(F. Poirion. Conférence au 1er colloque de mécanique aléatoire (annales du LCPC - décembre 1984) et article à paraître.

[7] F. Poirion, Lecture in the workshoop of Random Vibrations. (Paris june 1984) Published in the Annales of LCDC (december 1984).

Paul KREE
Laboratoire d'Analyse et Géométrie LA n°213
Université de Paris VI
Place Jussieu
75005 PARIS
FRANCE

E H LIEB

A lower bound on the Chandrasekhar mass for stellar collapse

INTRODUCTION

This article is a brief summary of some recent work with W. Thirring [1]; it also contains two conjectures about sharp constants in certain inequalities.

The basic quantity of interest is the following modified Schrödinger operator, for N particles, on $L^2(\mathbb{R}^{3N})$

$$H = \sum_{i=1}^{N} |p_i| - K \sum_{1 \le i \le j \le N} |x_i - x_j|^{-1} \qquad (1)$$

where $|p_i| \equiv (-\Delta_i)^{1/2}$. The usual Schrödinger operator would have $|p_i|^2 = -\Delta_i$ (actually $-\text{tr}^2\Delta_i/2m$) in place of $|p_i|$, and it is this replacement that is responsible for the "collapse" discussed herein. The use of $|p_i|$ is a crude attempt to make the Schrödinger equation relativistic (Actually, we should use $c[|p_i|^2 + m^2c^2]^{1/2}$ where c is the speed of light (which we take to be unity). But $|p| \le [|p|^2 + 1]^{1/2} \le |p| + 1$ so our H, and the correct H differ by a bounded operator which is irrelevant for the collapse question considered here).

The coupling constant is

$$K = Gm^2 > 0 \qquad (2)$$

where G is the gravitational constant.

The operator H has the following property : under the dilation $x_i \to \lambda x_i$ with $\lambda > 0$, $H \to H/\lambda$. Since this dilation is unitary, $\sigma = \text{spec}(H)$ is a subset of $\mathbb{R}$ that is invariant under $\mathbb{R} \to \lambda\mathbb{R}$, $\forall\lambda > 0$. Thus

$$E_0 \equiv \inf \text{spec}(H) = -\infty \text{ or } 0 \qquad (3)$$

The choice of $-\infty$ or 0 depends on K and N. The case $E_0 = -\infty$ has the physical meaning of *instability* or *collapse*, E_0 also depends on "statistics" which is defined as follows :

- $L^2(\mathbb{R}^{3N})$ is the N-fold tensor product of $L^2(\mathbb{R}^3)$ which has two (and more) subspaces that are invariant under H, namely the symmetric and the antisymmetric tensor products. Spec(H) will be different on the two spaces and we can ask for E_o on each. The symmetric case is called the *boson* case; the antisymmetric cas is the *fermion* case. (This is quite accurate; for the fermion case one should use the antisymmetric tensor product of $L^2(\mathbb{R}^3;C^2)$, but this refinement will be ignored here for simplicity). It is easy to prove using the positivity of $\exp(-|p|t)$ that E_o (on all of $L^2(\mathbb{R}^{3N})$) = E_o (boson) = E_o^b. The known fact about E_o are the following (with the superscripts b and f denoting the two cases) :

- there exists a critical $K_c = K_c(N)$ such that $E_o = -\infty$ (resp.0) for $K > K_c$(resp. $K < K_c$) .

- as $N \to \infty$:

$$C_b \geq N K_c^b \geq C_b' \tag{4.6}$$

$$C_f \geq N^{2/3} K_c^f \geq C_f' \tag{4.7}$$

for suitable constants C_b, etc... Presumably NK_c^b has a limit, C_b, as $N \to \infty$ (and similarly for $N(K_c^f)^{3/2}$). This is easy to prove in the boson case but a proof has not been given for the fermion case).

What are these sharp constants ?

In the sequel, conjectured values will be given.

If one takes the values (4) for K_c and takes m to be a proton mass, then the total critical mass, $M = Nm$, can be computed from $K = Gm^2$. With the upper bounds C_b, C_f, one finds that M is a little bigger than the mass of the sun in the fermion case and is about the size of a mountain in the boson case.

The fermion case is the most relevant physically. In the early days of quantum mechanics several authors [2]realized the existence of this critical mass, but it was Chandrasekhar who really took it seriously. Ultimately, his idea that sufficiently big stars would collapse when their "nuclear fuel" was burnt was accepted, and it is now the current theoretical basics for the formation of white awarfs and neutron stars; the earned him the Nobel prize en 1983. Chandrasekhar derived the upper bound C_f but until the work reported here [1], no one proved stability ($E_0 = 0$) when $K < (\text{const})N^{-2/3}$ The fact that the dependence of K_c on N is $N^{-2/3}$ (and not N^{-1}) is the difficult point about the fermion case.

2. THE BOSON CASE (UPPER BOUND)

To derive an upper bound for K_c we merely have to find some symmetric $\psi(x_1,\ldots,x_n)$ such that $(\psi,H\psi) < 0$. Take

$$\psi = \prod_{r=1}^{N} f(x_i) \tag{5}$$

with $f \in L^2(\mathbb{R}^3)$ and $\|f\|_2 = 1$. Then

$$(\psi,H\psi) = NA(f,C) \tag{6}$$

where

$$A(f,C) = (f,|p|f) - C\iint |f(x)|^2\ |f(y)|^2\ |x-y|^{-1}\ dx\ dy \tag{7}$$

$$C = \frac{K(N-1)}{2} \tag{8}$$

Thus in (4.b) :

$$C_b = 2\inf\{C|A(f,C) > 0,\ \forall f \text{ with } \|f\|_2 = 1\} \tag{9}$$

The *conjecture* is that the sharp constant $C_b(=\lim_{N\to\infty} N\,K_c^b)$ is given by (9). It is easy to prove the $C_b > 0$ since $\|\,|p|^{1/2} f\,\|_2 \geq (\text{const})\,\|f\|_3^2$ (Sobolev type inequality) and $\int |f(x)|^2\ |f(y)|^2\ |x-y|^{-1} \leq (\text{const})\ \|f\|_{12/5}^4$ (Hardy-Littlewood-Sobolev inequality).

3. THE BOSON CASE (LOWER BOUND).

First, consider the following operator on $L^2(\mathbb{R}^3)$:

$$h = |p| - \frac{z}{|x|}\ , \qquad z > 0 \tag{10}$$

h has the same property as H, namely $E_0 = \inf.\,\mathrm{spec}(h) = -\infty$ or 0 depending only on z. In the simple case the critical z_c is known. Clearly $h \geq 0$ if and only if the operator

$$\mathcal{H} = |x|^{-1/2}\, |p|^{-1}\, |x|^{-1/2}$$

satisfies $z\mathcal{H} \in I$. Now $|p|^{-1}$ has the explicit kernel

$$|p|^{-1}(x,y) = |x-y|^{-2} / (2\pi^2) \tag{11}$$

and thus $\mathcal{H}$ has the kernel

$$(x,y) = |x|^{-1/2}\, |x-y|^{-2}\, |y|^{-1/2}\, /(2\pi^2) \tag{12}$$

It is known that the norm of $\mathcal{H}$ satisfies

$$\|\mathcal{H}\| = \frac{\pi}{2} \tag{13}$$

(13) is a weighted Hardy-Littlewood - Sobolev inequality. A mapping f for (13) does *not* exist, i.e. there is no $f \in L^2$ such that $(f, \mathcal{H} f) = \frac{1}{2}\pi \|f\|_2$. The sharp constant for (10) is

$$z_c = \frac{2}{\pi} \tag{14}$$

and $h \geq 0$ iff $z \leq z_c$. For this and related matters see [3],[4].

Returning to H in (1) we can write

$$H = \sum_{i=1}^{N} \sum_{j \neq i} H_{ij} \tag{15}$$

$$H_{ij} = (N-1)^{-1} |p_i| - K|x_i - x_j|^{-1} \tag{16}$$

Using (14), we see that $H \geq 0$ if

$$(N-1)K \leq \frac{2}{\pi} \tag{17}$$

and thus

$$C_b' = \frac{2}{\pi}$$

in (4b).

4. THE FERMION CASE (UPPER BOUND)

As in section 2, we want to find an antisymmetric $\psi(x_1,\ldots,x_N)$ with $\|\psi\|_2 = 1$ such that $(\psi,H\psi) < 0$ when the left side of (4f) is violated. This is more difficult than in section 2.

Given any antisymmetric ψ one can define

$$\rho(x) = N \int |\psi(x,x_2,\ldots,x_N)|^2 \, dx_2 \ldots dx_N \tag{18}$$

so that

$$\int \rho(x)dx = N \quad \text{and} \quad \rho(x) \geq 0 \tag{19}$$

Let us define

$$T\psi = (\psi, \sum_{i=1}^{N} |p_i|\psi) \tag{20}$$

$$V\psi = (\psi, \sum_{i<j} |x_i-x_j|^{-1}\psi) \tag{21}$$

Physicists "know" (from semiclassical considerations) that if a reasonable nice ρ is given that satisfies (19), then there exists a ψ satisfying (18) and also

$$T\psi \leq \frac{3}{4} D_c \int \rho(x)^{4/3} dx \tag{22}$$

$$V\psi \geq \frac{1}{2} \int\int \rho(x)\, \rho(y)\, (x-y)^{-1} dx\, dy \tag{23}$$

apart from "unimportant terms" on the right sides of (22)(23). The constant D_c (here c denotes "classical") is

$$D_c = (6\pi^2)^{1/3} \tag{24}$$

Let us suppose that (22), (23) are strictly correct. Then all we have to do to obtain an upperbound on E_o is to minimize (with respect to ρ satisfying (19)) the functional

$$\mathcal{E}(\rho) = \frac{3}{4} D_c \int \rho^{4/3} - \frac{1}{2} K \iint \rho(x)\rho(y) \ |x-y|^{-1} \ dx \ dy \tag{25}$$

Let E_o^1 denote the infimum of $\mathcal{E}(\rho)$. It is easy to see that $E_o^1 = 0$ or $-\infty$ (since the two terms scale the same way under dilation) depending on K.

The minimization problem for $\mathcal{E}$ can be expressed another way. Consider the maximization problem

$$\mathbb{C} = \sup_g \|g\|_{4/3}^{-4/3} \ \|g\|_1^{-2/3} \iint g(x)g(y) \ |x-y|^{-1} dx \ dy \tag{26}$$

for all $g \in L^{4/3}(\mathbb{R}^3) \cap L^1(\mathbb{R}^3)$. It can be proved, [5] , that a maximizing g exists for (26) ; $g = f^3$ where f satisfies Emden's equation of order 3 This $g \in L^\infty$ and has compact support. The numerical value of $\mathbb{C}$ is

$$\mathbb{C} = 1.092 \tag{27}$$

Comparing (26) and (25) we see that $E_o^1 = 0$ if and only if

$$N^{2/3} K \le \frac{3}{4} \frac{D_c}{\mathbb{C}} = 2.677 \equiv C_f \tag{28}$$

The factor $N^{2/3}$ comes from the factor $\|g\|_1^{2/3}$ in (26) . The *conjecture* is that this C_f is the sharp constant for $\lim_{N\to\infty} N^{2/3} K_c^f$ (Compare (9) in section 2).

Thus, we can conclude that C_f in (28) can be used in (4f) provided we can justify (22), (23) with $\rho(x)$ being proportional to the maximizing g for (26) . The calculation (25) - (28) was done by the earlier workers; the construction of ψ satisfysing (22), (23) (with unimportant error terms) was first done in [1] . This construction uses the notion of "coherent states" and some difficult inequalities [5] that will not be discussed here.

5. THE FERMION CASE (LOWER BOUND)

Now we come to the heart of the matter - proving that $\inf \operatorname{spec}(H) = 0$ if $N^{2/3} K < C_f'$ 0 . As we saw in section 3, it is easy to find the weaker bound $NK < C_b'$.

First we decompose H in a manner reminiscent of (15). Fix positive integers L, M with $L+M = N$ (the optimum choice will turn out to be $M = O(N^{-1/11})$. Let Π denote a partition of $\{1,\ldots,N\}$ into 2 disjoint sets π_1 and π_2 of cardinality L and M. Let $\lambda,\alpha > 0$ satisfy

$$2\lambda(N-M)M - \alpha M(M-1) = K(N-M)(N-1) \tag{29}$$

Then H can be written

$$H = \frac{N}{2}\binom{N}{2}^{-1} \sum_{\Pi} \{ \sum_{i \in \pi_1} |p_i| - \lambda \sum_{i \in \pi_1} \sum_{j \in \pi_2} |x_i - x_j|^{-1} + $$

$$+ \alpha \sum_{j \in \pi_2} \sum_{j<k \in \pi_2} |x_j - x_k|^{-1} \} \tag{30}$$

If each operator in { } in (30) is ≥ 0 then clearly $H > 0$. Taking $\pi_1 = \{1,\ldots,L\}$, $\pi_2 = \{L+1,\ldots,M\}$ and denoting the second set of variables by y_j ($j = 1,\ldots,M$) - instead of x_j , $j = L+1,\ldots,N$ - we have to prove that

$$h_{L,M,\lambda} \equiv \sum_{i=1}^{L} |p_i| - \lambda \sum_{i=1}^{L} \sum_{j=1}^{M} |x_i - y_j|^{-1} \tag{31}$$

as an operator on the L fold antisymmetric tensor product of $L^2(\mathbb{R}^3)$, satisfies the operator inequality

$$h_{L,M,\lambda} \geq -\alpha w(g) \tag{32}$$

for every choice of $y_1,\ldots,y_M \in \mathbb{R}^3$ (distinct). In (32) $w(y)$ is the constant defined by

$$w(y) = \sum_{1\le i\le j\le M} |y_i - y_j|^{-1} \tag{33}$$

Fix the points $\{y\}$ and let $R_j > 0$ denote half the nearest neighbor distance from y_j, i.e.

$$2R_j = \min_{j\neq k} |y_j - y_k| \tag{34}$$

Let $V(x)$ be the potential in (31), i.e.

$$V(x) \equiv \sum_{j=1}^{M} |x - x_j|^{-1} \tag{35}$$

We choose some fixed t, $0 < t < 1$ and write

$$V(x) = V_+(x) + V_-(x) \tag{36}$$

where

$$V_+(x) = |x|^{-1} * G \tag{37}$$

$$G(x) = \sum_{j=1}^{M} g_j(x) \tag{38}$$

$$g_j(x) = (4\pi R_j^2 t^2)^{-1} \delta(|x - y_j| - tR_j) \tag{39}$$

Clearly, V_+ is the long range part of V and is the potential generated by δ masses spread over (disjoint) spheres of radius tR_j centered at the points y_j. V_-, the remainder, is the short range part.

Let B_j denote the ball of radius R_j centered at y_j; B_j will also be used to denote the characteristic function of this ball. These balls are disjoint. Obviously (by Newton's theorem)

$$V_-(x) = \sum_{j=1}^{M} B_j(x) |x - y_j|^{-1} \equiv U(x) \tag{40}$$

Next, choose $1 > \varepsilon > 0$ (the optimum choice will turn out to be $1-\varepsilon = O(N^{-1/11})$) ans split $h_{L,M,\lambda}$ into two pieces

$$h_{L,M,\lambda} \geq h_+ + h_- \tag{41}$$

$$h_+ = \sum_{i=1}^{L} \{\varepsilon|p_i| - \lambda V_+(x_i)\} \tag{42}$$

$$h_- = \sum_{i=1}^{L} \{(1-\varepsilon)\,|p_i| - \lambda U(x_i)\} \tag{43}$$

We shall show that $h_+ \geq -\alpha w(y)$ and $h_- \geq 0$ for suitable $\varepsilon, \lambda, \alpha$, K.

Lower bound for h_+ :

Given $\psi(x_1,\ldots,x_L)$ antisymmetric and with $\|\psi\|_2 = 1$, we want to show that

$$(\psi, h_+ \psi) \geq -\alpha w(y) \tag{44}$$

Define $\rho(x)$ as in (18) - with L and with $\int\rho = L$. There is a converse to (22), proved by Daubechies [6] :

$$T\psi \geq \frac{3}{4} D \int \rho(x)^{4/3}\,dx \tag{45}$$

for all ψ and with $D = 2.17$ (*Conjecture* : the sharp D in (45) is D_c in (34)). (Inequality (45) is an extension of the $\int \rho^{5/3}$ bound [7] for fermions with $p^2 = -\Delta$, instead of $|p|$.
Daubechies' proof uses the methods developed in [8]).

The second term in h_+, when inserted in (44), is precisely $-\lambda\int \rho V_+$. Thus,

$$(\psi, h_+ \psi) \geq \mathcal{F}(\rho) \tag{46}$$

with

$$\mathcal{F}(\rho) = \frac{3}{4} D\varepsilon \int \rho(x)^{4/3} - \lambda \int \rho(x) V_+(x)\,dx \tag{47}$$

A lower bound is obtained by minimizing $\mathcal{F}$ with respect to ρ(with $\rho \geq 0$, but without the condition $\int\rho = L$) :

$$(\psi,h_+ \psi) \geq -\frac{1}{4}(\varepsilon D)^{-3} \lambda^4 \int V_+(x)^4 \, dx \tag{48}$$

From (37) and the Schwarz inequality

$$V_+(x)^2 \leq \left(\int G\right) \{|x|^2 * G\}(x) = M\{|x|^{-2} * G\}(x) \tag{49}$$

Since $|x|^{-2} * |x^{-2}| = \pi^3 |x|^{-1}$

$$\int V_+^4 \leq \pi^3 M^2 \iint G(x)G(y) \, |x-y|^{-1} \, dx \, dy \tag{50}$$

The integral in (50) is (by Newton's theorem)

$$2W(y) + \sum_{j=1}^{M} \iint g_j(x) \, g_j(y) \, |x-y|^{-1} \, dx \, dy = \tag{51}$$

$$= 2W(y) + \sum_{j=1}^{M} (R_j t)^{-1}$$

But $W \geq \frac{1}{4} \sum_{j=1}^{M} R_j^{-1}$, and thus

$$h_+ \geq -\frac{1}{2}(\varepsilon D)^{-3} \pi^3 M^2 \lambda^4 (1 + \tfrac{2}{t}) W(y) \tag{52}$$

This is precisely the kind of bound we need (cf. (32)) , provided we can choose $L,\alpha,\varepsilon,\lambda,t$ such that

$$\pi^3 M^2 \lambda^4 (1 + \tfrac{2}{t}) \leq 2(\varepsilon D)^3 \alpha \tag{53}$$

Lower bound for h_-.

Obviously, it suffices to prove that the simple particle operator on $L^2(\mathbb{R}^3)$

$$\tilde{h} = (1-\varepsilon)|p| - \lambda U(x) \tag{54}$$

satisfies $\tilde{h} \geq 0$.

In other words, we forego the advantage of using Fermi statistics. As in (10) - (13), we have to show that with

$$\mathcal{K} = U(x)^{1/2} \; |p|^{-1} \; U(x)^{1/2} \tag{55}$$

then

$$||\mathcal{K}|| \leq \frac{1-\varepsilon}{\lambda} \tag{56}$$

With $w_j(x) = |x-y_j|^{-1/2}$ (and recalling that the B_j are disjoint)

$$\mathcal{K} = \sum_{j=1}^{M} \sum_{k=1}^{M} K_{jk} \tag{57}$$

$$2\pi^2 K_{jk}(x,y) = B_j(x) w_j(x) \; |x-y|^{-2} \; w_k(y) B_k(y) \tag{58}$$

Take $f \in L^2(\mathbb{R}^3)$ and first consider the diagonal terms $j = k$ (57) in $(f, \mathcal{K} f)$. Using (12), (13)

$$\sum_{j=1}^{M} (f, K_{jj} f) \leq \frac{\pi}{2} \sum_{j=1}^{M} (B_j f, B_j f) \leq \frac{\pi}{2}(f,f) \tag{59}$$

Next, for the $j \neq k$ terms we use the fact that $|x-y| \geq (1-t) \; |y_j - y_k|$ when $x \in B_j$ and $y \in B_k$. Thus

$$\sum_{j \neq k} (f, K_{jk} f) \leq \sum_{j \neq k} n_j n_k |y_j - y_k|^{-2} (1-t)^{-2}/(2\pi)^2 \tag{60}$$

with

$$n_j = \int f \, B_j \, w_j \leq (2\pi)^{1/2} \, R_j \, t \int f^2 \, B_j \tag{61}$$

It is left as an exercise (cf. [1]) to show that the sum in (60) satisfies (for any n_j)

$$\sum_{j \neq k} n_j n_k |y_j - y_k|^{-2} \leq \left(\frac{3\pi}{2}\right)^{2/3} \left(\sum_j n_j^{3/2} R_j^{-3/2} \right)^{4/3} \tag{62}$$

(This requires using the sharp constant in the Hardy-Littlewood - Sobolev inequality [4]; (62) resembles the ordinary HLS inequality except that it is a sum instead of an integral).

Combining the previous estimates,

$$|| \ || \leq \frac{\pi}{2} + t^2(1-t)^{-2} \ (9M)^{1/3} \ (4\pi)^{-1/3} \tag{63}$$

and we shall require that the right side of (63) $\leq$ $1-\varepsilon/\lambda$.

<u>CONCLUSION</u> : The parameters $L,\lambda,\alpha,\varepsilon,t$ have to be chosen to satisfy (29), (53), (56), (61). A simple algebraic exercise shows that this can be done if

$$N^{2/3}K \ \leq 3^{2/3} \ D(2\pi)^{-1} \ (1+O(N^{-1/11})) \tag{64}$$

With Daubechies'bound [6], $D \geq 2.17$, we have that (4f) hold with

$$C'_f = 0.718 \tag{65}$$

This should be compared with (28).

6. <u>EXTENSIONS</u>

The hamiltonian H in (1) contains only the kinetic and gravitational energies; moreover it is concerned with only one kind of particle (all bosons or all fermions). For physical applications one wants to consider several kinds of particles (with different masses) and also Coulomb interactions, namely

$$H' = \sum_{i=1}^{N} |p_i| + \sum_{1\leq i\leq j\leq N} (e_i e_j - Gm_i m_j) \ |x_i - x_j|^{-1} \tag{66}$$

Here e_i are the elastic charges (positive or negative) and $m_i > 0$ are the masses.

Since G is tiny, $|e_i e_j| >> Gm_i \ m_j$ for all i,j.

One might think that the Coulomb forces dominate. This is not so. There is concellation since the e_i can have two signs. In [1] , it is shown that the gravitational forces are primarily what determines the collapse. The Coulomb forces play a minor rôle.

BIBLIOGRAPHY

[1] E.H. Lieb, W.E. Thirring. Gravitational collapse in quantum mechanics and relativislic kinetic energy. Annals of Physics (to appear)

[2] R. Fowler. Monthly Notices. 87 (114), 1926. I. Frenkel. Z. Für Physik. 50 (234), 1928. W. Anderson. Z für Physik. 56 (851), 1929.
E. Stoner. Phil. Mag. 9 (944), 1930.
S. Chandrasekhar. Phil. Mag. 11 (592), 1931.
L. Landau. Phys. Z.d. Sovietunion. 1 (285), 1932.

[3] I. Daubechies, E.H. Lieb. Comm. Math. Phys. 90 (497), 1983.

[4] E.H. Lieb. Annals of Math. 118 (349), 1983.

[5] E.H. Lieb, S. Oxford. Int. J. Quant. Chem. 19 (427), 1981.

[6] I. Daubechies. Comm. Math. Phys. 90 (511), 1983

[7] E.H. Lieb, W.E. Thirring. Phys. Rev. Lett. 35 (687), 1975; Errata ibid. 35 (111 b), 1975). See also : studies in Mathematical Physics : Essays in Honor of V. Bargmann. E. Lieb, B. Simon, A. Wightman eds, Princeton University Press (1976), 269-303.

[8] E.H. Lieb. Bull. Amer. Math. Soc, 82 (751), 1976. The details appear in Proc. Amer. Math. Soc. Symp in Pure Math. R. Osserman and A. Weinstein eds, 36 (1980), 241-252.

Work partially supported by U.S. National Science Foundation grand PHY 81-16101- A02.

Elliott H. LIEB
Departments of Mathematics and Physics
Princeton University
P.O. Box 708
PRINCETON
N.J. 08544
U.S.A.

Y MEYER

Etude d'un modèle mathématique issu du controle des structures spatiales déformables

INTRODUCTION.

Le contrôle des structures spatiales flexibles a conduit J.L. Lions à étudier le modèle simplifié suivant, appelé équation d'état.

On désigne par $\Omega \subset \mathbb{R}^3$ un ouvert borné régulier, par $T > 0$ un réel positif. On pose $Q = \Omega \times]0,T[$, $\Sigma = \partial\Omega \times]0,T[$ et l'on étudie, pour $b \in \Omega$ donné et $v(t) \in L^2[0,T]$ donnée, le système

$$\left\{\begin{array}{l} \dfrac{\partial^2 y}{\partial t^2} - \Delta y = v(t)\delta(x-b) \text{ dans } Q;\ \delta(x-b) \text{ est la masse de Dirac en } b \\ y(x,0) = \dfrac{\partial y}{\partial t}(x,0) = 0 \quad \text{dans } \Omega \\ y = 0 \text{ sur } \Sigma. \end{array}\right. \tag{1}$$

On désigne par $y(x,t,v) = y(v)$ la solution de (1) dans un espace à préciser et on cherche à minimiser, pour un profil donné $z_d(x) \in L^2(\Omega)$, la fonction coût suivant $J(v) =$

$$\int_\Omega [y(x,T;v) - z_d(x)]^2 dx + N\int_0^T v^2(t)dt \tag{2}$$

sur un ensemble convexe fermé de contrôles admissibles $v \in \mathcal{U}$.

Toutes les fonctions sont à valeus réelles et $N > 0$ est donné.

J.L. Lions remarque avec force que ce problème de contrôle optimal pose un problème préalable encore ouvert : pour v donné dans $L^2(0,T)$, la fonction $x \to y(x,T;v)$ est-elle dans $L^2(\Omega)$?

Nous allons montrer qu'il en est effectivement ainsi et préciser l'espace fonctionnel où l'on cherche les solutions.

Théorème 1. *Le système* (1) *possède, pour tout contrôle* v(t) × $L^2[0,T]$, *une et une seule solution* y ∈ $\mathcal{E}$ *où* $\mathcal{E}$ *est l'espace fonctionnel défini par les trois conditions suivantes*

$$\begin{cases} y(.,t) \in C([0,T]; L^2(\Omega)) \\ \frac{\partial y}{\partial t}(.,t) \in C([0,T]; H^{-1}(\Omega)) \\ \frac{\partial^2 y}{\partial t^2}(.,t) \in L^2([0,T]; H^{-2}(\Omega)). \end{cases} \quad (3)$$

De plus on a, pour une constante C(Ω) *ne dépendant que de la géométrie de* Ω *et en supposant* $\|v\|_{L^2[0,T]} \leq 1$,

$$\sup_{0\leq t\leq T} \|y(.,t)\|_{L^2(\Omega)} \leq C(\Omega)\ \sqrt{1+T}\ \delta_b^{-1/2} \quad (4)$$

où δ_b = distance(b,∂Ω).

La démonstration de ce résultat utilise trois ingrédients (a) l'algorithme de dualité présenté par J.L. Lions dans [0] et permettant de reporter le problème sur l'étude de *l'état adjoint.*

(b) des techniques classiques sur les sommes trigonométriques apériodiques.

(c) un théorème de Hörmander, raffiné par J. Brünig et Pham The Lai sur le comportement asymptotique de la fonction spectrale associée à un ouvert borné régulier.

1. L'ALGORITHME DE DUALITE.

L'état adjoint p(x,t) : Q → ℝ est solution de

$$\begin{cases} \frac{\partial^2 p}{\partial t^2} - \Delta p = 0 \text{ dans } Q \\ p(x,T) = 0 \text{ dans } \Omega \text{ et } \frac{\partial p}{\partial t}(x,T) = g \in L^2(\Omega) \\ p = 0 \text{ sur } \Sigma \end{cases} \quad (1.1)$$

Un calcul très simple montre alors que

$$\int_Q \left(\frac{\partial^2 y}{\partial t^2} - \Delta y\right)p(x,t)\,dxdt = \int_0^T v(t)p(b,t)\,dt = - \int_\Omega g(x)y(x,T;v)\,dx.$$

Nous voulons prouver l'inégalité $\|y(x,T)\|_{L^2(\Omega)} \leq C\|v\|_{L^2[0,T]}$. Il suffit pour cela de démontrer que

$$\|p(b,t)\|_{L^2[0,T]} \leq C\|g\|_{L^2(\Omega)}.$$

C'est ce que nous allons faire maintenant.

2. ESTIMATIONS SUR LES SOMMES TRIGONOMETRIQUES APERIODIQUES.

L'ensemble S des fréquences est une suite strictement croissante $s(j)$, $j \in \mathbb{Z}$, de nombres réels, telle que $s(-j) = - s(j)$. On appelle $\omega(j) \geq 0$, $j \in \mathbb{Z}$, un poids soumis à la condition $\omega(-j) = \omega(j)$, $j \in \mathbb{Z}$. Le problème que nous nous proposons de résoudre est l'existence d'un $T > 0$ et d'une constante $C(T) > 0$ tels que, pour toute suite $\alpha(j) \in \ell^2(\mathbb{Z})$, on ait

$$\int_0^T \Big|\sum_{-\infty}^{+\infty}\alpha(j)\omega(j)\exp i\, s(j)t\Big|^2 dt \leq C(T) \sum_{-\infty}^{+\infty} |\alpha(j)|^2 \tag{2.1}$$

Nous allons résoudre complètement ce problème (théorème 2 ci-dessous) grâce à une suite de lemmes.

Lemme 1. *Soit* $d\mu(\xi)$ *une mesure de Radon* ≥ 0 *sur* $\mathbb{R}$ *et soit* $T > 0$. *Une condition nécessaire et suffisante pour qu'il existe une constante* $C(T) > 0$ *telle que*

$$\int_{\mathbb{R}} |\hat{f}|^2 d\mu \leq C\int_0^T |f(x)|^2 dx \tag{2.2}$$

pour toute fonction $f \in L^2(\mathbb{R})$ *nulle hors de* $[0,T]$ *est que* $\sup_{k\in\mathbb{Z}} \mu([k,k+1]) \leq C'$.

Cette dernière condition montre que $T > 0$ ne joue aucun rôle si ce n'est dans le choix optimal de la constante figurant dans (2.2) que l'on peut choisir $\leq C''(1+T)$.

Montrons d'abord que (2.2) implique la propriété géométrique de μ.

Pour cela partons d'une fonction $\phi \in L^2(\mathbb{R})$, portée par $[0,T]$ et dont la transformée de Fourier ne s'annule pas sur $[0,1]$ et formons $f(x) = e^{ikx}\phi(x)$. Alors $|\hat{f}(\xi)| \geq \gamma > 0$ sur $[k,k+1]$ où $\gamma = \inf_{[0,1]} |\hat{\phi}|$. Il suffit alors d'écrire (2.2) pour obtenir la conclusion désirée.

En sens inverse, nous utiliserons le lemme suivant

Lemme 2. *On a*

$$\sum_{-\infty}^{+\infty} \{ \sup_{[k,k+1]} |\hat{f}|^2 \} \leq 10^6 (1+T) \int_0^T |f|^2 dx \tag{2.3}$$

pour toute fonction $f \in L^2(\mathbb{R})$ *portée par* $[0,T]$.

Remarque. Nous ne savons pas si 10^6 est optimal. En tout cas $1+T$ l'est pour les grandes valeurs de T comme on le voit en prenant $f(x) = 1$ sur $[0,T]$, $f = 0$ ailleurs.

Appelons $\phi(x) \in \mathcal{D}(\mathbb{R})$ une fonction égale à 1 sur $[0,1]$ et posons $\phi_T(x) = \phi(\frac{x}{T})$. On définit successivement $p_T(\xi) = \hat{\phi}_T(\xi) = T\hat{\phi}(T\xi)$, $C_1 = \|p_T\|_1$ et $q_T(\xi) = \sup_{[\xi-1,\xi+1]} |p_T(\xi)|$. Un calcul facile donne $\|q_T\|_1 \leq C_2(1+T)$.

Revenons à f. On a $f = f\phi_T$ et donc $\hat{f}=\hat{f} * p_T$ ce qui entraîne (Cauchy-Schwarz), $|\hat{f}(\xi)|^2 \leq C_1 |\hat{f}|^2 * |p_T|$. Ensuite , grâce à la définition de q_T,

$$\sup_{[k,k+1]} |\hat{f}|^2 \leq C_1 \inf_{[k,k+1]} \{|\hat{f}|^2 * q_T\} \leq C_1 \int_k^{k+1} \{|\hat{f}|^2 * q_T\} d\xi.$$

Finalement $\displaystyle\sum_{-\infty}^{+\infty} \sup_{[k,k+1]} |\hat{f}|^2 \leq C_1 \int_{-\infty}^{+\infty} |\hat{f}|^2 * q_T d\xi \leq C_1 C_2 (1+T) \|f\|_2^2$.

Revenons au lemme 1.

On a

$$\int_{\mathbb{R}} |\hat{f}|^2 d\mu = \sum_{-\infty}^{+\infty} \int_k^{k+1} |\hat{f}|^2 d\mu \leq \sum_{-\infty}^{+\infty} \sup_{[k,k+1]} |\hat{f}|^2 \mu[k,k+1] \leq$$

$$C'(1+T) \int_0^T |f(x)|^2 dx.$$

Théorème 2. *Une condition nécessaire et suffisante pour que* (2.1) *ait lieu est l'existence d'une constante* C' *telle que*

$$\sup_{k\in \mathbb{N}} \sum_{k\leq s(j)<k+1} \omega^2(j) \leq C'$$

Cette condition est indépendante de $T > 0$ *et l'on peut alors choisir* $C(T) \leq C'' \sqrt{1+T}$.

De façon usuelle, nous linéarisons le problème en calculant la norme $L^2[0,T]$ de $S(t) = \sum_{-\infty}^{+\infty} \alpha(j)\omega(j)\exp i\, s(j)t$ comme $\sup|\langle s,u\rangle|$ où $u \in L^2(\mathbb{R})$ est nulle hors de $[0,T]$ et vérifie $\|u\|_2 \leq 1$.

Moyennant un changement innocent de j en -j, (2.1) est équivalent à

$$\left|\sum_{-\infty}^{+\infty} \alpha(j)\omega(j)\hat{u}(s(j))\right| \leq C(T) \tag{2.4}$$

lorsque $\sum_{-\infty}^{+\infty} |\alpha(j)|^2 \leq 1$ et $\|u\|_2 \leq 1$.

En prenant la borne supérieure par rapport aux $\alpha(j)$, ceci est encore équivalent à

$$\sum \omega^2(j)|\hat{u}(s(j))|^2 \leq C^2(T). \tag{2.5}$$

On appelle finalement $d\mu$ la somme des masses ponctuelles ω_j^2 placées en s(j) et l'on applique le lemme 1.

Dans les applications que nous aurons à traiter, $\alpha_j\omega_j$ se présente, en fait, sous la forme $\sum_{k\in E(j)} \alpha(j,k)\omega(j,k)$ où E(j) est, pour chaque $j \in \mathbb{Z}$, un ensemble fini, les $\omega(j,k)$ sont des nombres complexes donnés et où

$\sum\sum|\alpha(j,k)|^2 < +\infty$. On se ramène immédiatement au cas précédent en posant $\tilde{\omega}(j) = (\sum_{k\in E(j)} |\omega(j,k)|^2)^{1/2}$ et $\tilde{\alpha}(j) = \frac{1}{\tilde{\omega}(j)} \sum_{k\in E(j)} \alpha(j,k)\omega(j,k)$. On a alors $\sum|\tilde{\alpha}(j)|^2 \le \sum\sum|\alpha(j,k)|^2$ et la condition nécessaire et suffisante pour obtenir (2.1) est l'existence d'une constante C telle que, uniformément en $m \in \mathbb{N}$, on ait

$$\sum_{\substack{k\in E(j) \\ m\le s(j)<m+1}} \sum |\omega(j,k)|^2 \le C. \tag{2.6}$$

3. ESTIMATIONS SUR LA FONCTION SPECTRALE.

Soit $\Delta : H_0^1(\Omega) \to H^{-1}(\Omega)$ l'isomorphisme usuel (donné par Lax-Milgram), posons $V = \Delta^{-1}(L^2(\Omega))$ et notons $\Delta : V \to L^2(\Omega)$ $V \subset H_0^1(\Omega)$ l'opérateur auto-adjoint de domaine maximal. On appelle $-\lambda_j$ les valeurs propres et $F_j \subset L^2(\Omega)$ les sous-espaces propres correspondants. Soit, pour chaque j, $f_{j,k}(x)$ une base orthonormée (pour la structure hilbertienne de $L^2(\Omega)$) de F_j.

On pose $e(x,y,\lambda) = \sum\sum_{\lambda_j\le\lambda} f_{j,k}(x)\overline{f_{j,k}}(y)$; $e(x,y,\lambda)$ est le noyau de l'opérateur P_λ de projection orthogonale (au sens de l'espace de Hilbert $L^2(\Omega)$) sur la somme des sous-espaces propres F_j tels que $\lambda_j \le \lambda$.

On forme $r(x,x,\lambda) = e(x,x,\lambda) - \frac{1}{2\pi^2}\lambda^{3/2}$, $x \in \Omega$ et l'on a (Hörmander, The spectral function of an elliptic operator, Acta. Math. 121 (1968) 193-218) $r(x,x,\lambda) = O(\lambda)$ uniformément sur tout compact $K \subset \Omega$. Plus précisément J. Brünig a prouvé que $r(x,x,\lambda) = O(\lambda\delta_x^{-1})$ où δ_x est la distance de x à $\partial\Omega$.

Nous pouvons terminer la preuve du théorème 1 ou, du moins, la démonstration de la partie essentielle. A savoir l'inégalité $(\int_0^T |p(b,t)|^2 dt)^{1/2} \le C(\Omega)(1+T)^{1/2} \delta_b^{-1/2} \|g\|_2$ sur l'état adjoint.

On écrit $g(x) = \sum\sum\alpha(j,k)f_{j,k}(x)$ où $\sum\sum|\alpha(j,k)|^2 = \|g\|_2^2$. On a alors (en échangeant les rôles de 0 et de T dans l'équation adjointe)

$$p(x,t) = \sum\sum\alpha(j,k)f_{j,k}(x)\,\frac{\sin\sqrt{\lambda_j}\,t}{\sqrt{\lambda_j}}\,.$$

L'inégalité désirée est, compte tenu du théorème 2, équivalente à

$$\sum\sum_{m\leq\sqrt{\lambda}_j<m+1} |f_{j,k}(b)|^2 \lambda_j^{-1} \leq C' \tag{3.1}$$

ou encore à $\sum\sum_{m\leq\sqrt{\lambda}_j<m+1} |f_{j,k}(b)|^2 \leq C'm^2$.

Mais cette dernière inégalité s'écrit aussi

$$e(b,b,m+1) - e(b,b,m) \leq C'm^2$$

et découle du théorème d'Hörmander.

Nous voulons maintenant démontrer que $y'(T) \in H^{-1}(\Omega)$. On considère, pour ce faire, un problème adjoint différent où $p(T) \in H_0^1(\Omega)$ et $p'(T) = 0$. Cela donne (après échange des rôles de T et de 0),

$$p(b,t) = \sum\sum\alpha(j,k)\cos\sqrt{\lambda}_j t\; f_{j,k}(b)$$

où $\sum\sum\alpha(j,k)f_{j,k}(x) \in H_0^1(\Omega)$ ce qui donne

$$\sum\sum|\alpha(j,k)|^2\lambda_j < +\infty.$$

Ceci nous amène à poser $\beta(j,k) = \lambda_j^{1/2}\alpha(j,k)$. On a $\sum\sum|\beta(j,k)|^2 < +\infty$ et

$$p(b,t) = \sum\sum\beta(j,k)\,\frac{\cos\sqrt{\lambda}_j t}{\sqrt{\lambda_j}}\, f_{j,k}(b).$$

Le calcul dr $(\int_0^T |p(b,t)|^2 dt)$ se fait en appliquant le théorème 2.

4. AUTRES FONCTIONS DE COUT ET AUTRE PREUVE DU THEOREME 1.

On peut considérer, pour le système (1) d'équation d'état, d'autres fonctions de coût. Par exemple

$$J(v) = \|y(T,v) - z_d^0\|^2_{H_o^1(\Omega)} + \|y'(T,v) - z_d^1\|^2_{L^2(\Omega)} + N\|v\|^2_{L^2[0,T]}$$

où $z_d = \{z_d^0, z_d^1\} \in H_o^1(\Omega) \times L^2(\Omega)$.

J.L. Lions pose le problème de caractériser les $v \in L^2[0,T]$ tels que $y(T,v) \in H_o^1(\Omega)$ et $y'(T,v) \in L^2(\Omega)$.

En utilisant les méthodes présentées ci-dessus, on montre facilement que $v \in H_o^1[0,T]$ entraîne ces deux propriétés.

Pour finir, nous allons donner une preuve directe du théorème 1 (sans faire appel à l'état dual).

Cherchons la solution $y(x,t)$ de (1) sous la forme $y(x,t) = \sum\sum \gamma_{j,k}(t) f_{j,k}(x)$ où $f_{j,k} \in H_o^1(\Omega)$ sont les fonctions propres normalisées de $\Delta : H_o^1(\Omega) \to H^{-1}(\Omega)$ et associées aux valeurs propres $-\lambda_j$.

On a, du moins formellement,

$$\sum\sum(\gamma''_{j,k}(t) + \lambda_j \gamma_{j,k}(t)) f_{j,k}(x) = v(t)\delta(x-b) \tag{4.1}$$

avec $\gamma_{j,k}(0) = \gamma'_{j,k}(0) = 0$.

En effectuant le produit scalaire de (4.1) avec chaque $f_{j,k}$ dans $L^2(\Omega)$, il vient

$$\gamma''_{j,k}(t) + \lambda_j \gamma_{j,k}(t) = v(t)\overline{f_{j,k}}(b) \tag{4.2}$$

avec $\gamma_{j,k}(0) = \gamma'_{j,k}(0) = 0$.

On a donc

$$\gamma_{j,k}(t) = \frac{1}{\sqrt{\lambda_j}} \int_0^t \sin(\sqrt{\lambda_j}(t-s))v(s)ds$$

et l'on cherche à prouver que $\sum\sum \gamma_{j,k}(T) f_{j,k} \in L^2(\Omega)$. Cela conduit à la condition nécessaire et suffisante $\sum\sum |\gamma_{j,k}(T)|^2 < +\infty$. Or

$$\gamma_{j,k}(T) = \frac{1}{\sqrt{\lambda_j}} \int_0^T \sin(\sqrt{\lambda_j}(T-t)) v(t) dt =$$

$$\frac{1}{2i\sqrt{\lambda_j}} (e^{i\sqrt{\lambda_j}T} \hat{v}(\sqrt{\lambda_j}) - e^{-i\sqrt{\lambda_j}T} \hat{v}(-\sqrt{\lambda_j})) \text{ et donc}$$

$$|\gamma_{j,k}(T)|^2 \leq \frac{1}{2\lambda_j} (|\hat{v}(\sqrt{\lambda_j})|^2 + |\hat{v}(-\sqrt{\lambda_j})|^2).$$

Ceci permet d'appliquer le théorème 2.

REFERENCES.

[1] L. Hörmander, The spectral function of an elliptic operator. Acta Math. 121 (1968) 193-218.

[2] J.L. Lions, Sur le contrôle ponctuel de systèmes hyperboliques. Séminaire Goulaouic-Schwartz du 28 Février 1984.

[3] Pham The Lai, Meilleures estimations asymptotiques des restes de la fonction spectrale et des valeurs propres relatifs au laplacien. Math. Scand. 48 (1981) 5-38.

Nous disposons, à l'heure actuelle, de deux nouvelles démonstrations des résultats précédents, obtenues successivement par L. Nirenberg et J.L. Lions.

Yves MEYER

Ecole Polytechnique
Département de Mathématiques

92118 PALAISEAU CEDEX

FRANCE

O PIRONNEAU

Transport de microstructures

1. LE PROBLEME MODELE.

Soit w^o une fonction régulière de $R^n \times R^n$ dans R^n à support compact et u^o une fonction régulière de R^n dans R^n. On cherche la limite lorsque $\varepsilon \to 0$ de $u^\varepsilon(x,t)$ solution de

$$\begin{cases} u_t + u\nabla u + \nabla p - \mu\varepsilon^2 \Delta u = 0, \quad \nabla.u = 0 \text{ dans } R^n \times]0,T[\\ u(x,0) = u^o(x) + w^o(x,\frac{x}{\varepsilon}) \end{cases} \qquad (1)$$

Remarque. L'écriture (1.b) n'a de sens que si $w^o(x,y)$ est de moyenne nulle y sur $Y =]0,1[^3$ (sinon comment définir u^o et w^o à partir de $u(x,0)$?). On note par < > l'opérateur de moyenne en y sur Y. L'analyse s'étend au cas où w^o est une variable aléatoire stationnaire en y (l'opérateur de moyenne < > est alors l'espérance). Ici on se limitera, pour les calculs, au cas périodique.

(H.1) Hypothèse : Le calcul qui suit suppose que u^ε est unique et régulier.

2. INTERPRETATION PHYSIQUE.

Le système (1) modélise les écoulements incompressibles à grand nombre de Reynolds. Les conditions initiales font apparaître l'écoulement moyen u^o et les "turbulences" $w(x,\frac{x}{\varepsilon})$. Savoir séparer les grandes échelles (u^o) des petites (w^o) n'est pas toujours possible ; pratiquement cela implique que le spectre d'énergie est nul ou petit au voisinage de $1/\varepsilon$; ce phénomène de "ségrégation" [1] existe pour les écoulements météorologiques à grande échelle (les petites turbulences sont 3-D alors que les grands tourbillons dans la couche planétaire sont 2-D, entre les 2 il y a un "vide spectral") et est plus ou moins vérifié dans les écoulements derrière un obstacle (aile d'avion) ou les jets.

L'écriture (1) contient implicitement une autre hypothèse : le spectre d'énergie des petites structures est d'ordre 1 en amplitude : cette hypothèse peut être relaxée sans difficulté [2]. Enfin (1) contient une 3ème hypothèse car la viscosité $\nu^\varepsilon = \mu\varepsilon^2$ est reliée à la taille ε des petites échelles. Si $\nu^\varepsilon \ll \mu\varepsilon^2$ le calcul qui va suivre reste valable [3] mais si $\nu^\varepsilon \gg \mu\varepsilon^2$ il semble qu'il y ait de sérieuse difficultés.

3. ANZATS.

Pour trouver un Anzats pour u^ε on remarque [4][5] que l'effet dominant de u^0 sur w sera un effet de convection. On note donc a(x,t) la coordonnée Lagrangienne du problème ($a \simeq x + ut$, si t est petit) et on pose :

$$\begin{cases} u^\varepsilon(x,t) = u(x,t) + w(y,x,t) + \varepsilon u^1(y,x,t) + \varepsilon^2(\ldots)\Big|_{y = \frac{a(x,t)}{\varepsilon}} \\ p^\varepsilon(x,t) = p(x,t) + \pi(y,x,t) + \varepsilon p^1(y,x,t) + \varepsilon^2(\ldots)\Big|_{y = \frac{a(x,t)}{\varepsilon}} \end{cases} \quad (2)$$

avec $w, u^1, \pi, p^1 \ldots$ Y-périodique en y et w, π de moyenne nulle.

Remarque. On pourrait introduire une échelle de temps rapide $\tau = t/\varepsilon^2$ mais ceci peut être évité moyennant une hypothèse sur w^0 ($w^0 \nabla w^0 + \nabla\pi = 0$). L'absence d'une telle échelle correspond à "l'hypothèse de Taylor" qui considère les oscillations en temps essentiellement dues à la convection des oscillations en espace.

Reportées dans (1), les équations (2) et (3) donnent :

$$0 = \frac{1}{\varepsilon} [a_t + (u+w)\nabla a].\nabla_y w + \nabla a \nabla_y \pi$$

$$+ \varepsilon^0 [a_t + (u+w).\nabla a].\nabla_y u^1 + (u^1.\nabla a).\nabla_y w + \nabla a \nabla_y p^1$$

$$+ (u+w)_t + (u+w)\nabla(u+w) + \nabla(p+\pi) - \mu \nabla a^T \nabla a : \nabla_y \nabla_y w]$$

$$+ \varepsilon\{[a_t + (u+w).\nabla a].\nabla_y u^2 + (u^2.\nabla a).\nabla_y w + \nabla a \nabla_y p^2 \qquad (3)$$

$$+ (u^1.\nabla a).\nabla_y u^1 + u^1_t + (u+w)\nabla u^1 + u^1 \nabla(u+w) + \nabla p^1 - \mu \nabla a^T \nabla a : \nabla_y \nabla_y u^1$$

$$- 2\mu \nabla a : \nabla_y \nabla w\} + \varepsilon^2(\ldots)$$

$$0 = \frac{1}{\varepsilon} \nabla_y.(\nabla a^T w) + \varepsilon^0 [\nabla_y.(\nabla a^T u^1) + \nabla.w] + \varepsilon[\nabla_y.(\nabla a^T u^2) + \nabla.u^1] + \varepsilon^2(\ldots) \qquad (4)$$

4. CASCADE D'EQUATIONS.

4.1. Annulation des termes en ε^{-1} :

$$\begin{cases} [a_t + (u+w).\nabla a].\nabla_y w + \nabla a \nabla_y \pi = 0 \\ \nabla_y.(\nabla a^T w) = 0 \end{cases} \qquad (5)$$

En posant

$$\tilde{w} = [a_t + (u+w).\nabla a] \left(\tfrac{1}{2}\langle w^2 \rangle\right)^{-1/2} \qquad (6)$$

On écrit (5)

$$\begin{cases} \tilde{w}.\nabla_y \tilde{w} + C \nabla_y \pi = 0 \\ \nabla_y.\tilde{w} = 0, \ \tfrac{1}{2}\langle \tilde{w}.C^{-1}\tilde{w} \rangle = 1, \ \langle \tilde{w} \rangle = 0 \end{cases} \qquad (7)$$

où

$$C = \nabla a^T \nabla a$$

(H.2) Hypothèse : A hélicité totale donnée

$$<\nabla a^{-T}\,\tilde{w}.(\nabla a \nabla) \times \nabla a^{-T}\,\tilde{w} = r, \qquad (8)$$

le problème (7) admet une solution isolée continuement différentiable par rapport à $\{C,r\}$ et égale à $w^o(\frac{1}{2}<w^{o^2}>)^{-1/2}$ si $C = I$ (i.e. $t=0$).

Remarque. On connait au moins une solution analytique :

$$\tilde{w} = \bar{b}e^{-iky} + be^{iky},\ b.k = \bar{b}.k = 0$$

satisfaisant l'hypothèse pour des w^o, particuliers. Numériquement on a obtenu des solutions pour $r=0$ pour, semble-t-il, tout w^o vérifiant (7) avec $c = I$.

Donc $\tilde{w}$ est déterminé par (7)-(8) et si en moyenne (6) on obtient, compte tenu de $<w> = 0$:

$$a_t + u\nabla a = 0 \qquad (9)$$

$$w = \sqrt{q}\ \nabla a^{-T}\,\tilde{w} \qquad (10)$$

$$q = \frac{1}{2}<w^2> \qquad (11)$$

Les quantités $q(x,t)$ et $r(x,t)$ seront déterminées ultérieurement.

4.2. Annulation des termes en ε^o.

Compte tenu de (9) l'annulation des termes d'ordre 0 donne un système du type

$$\begin{cases} \tilde{w}\nabla_y u^1 + (\nabla a^T u^1)\nabla_y w + \nabla a \nabla_y p^1 + f = 0 \\ \\ \nabla_y.(\nabla a^T u^1) = -\nabla.w \end{cases} \tag{12}$$

On pose donc

$$\tilde{u}^1 = \nabla a^T u^1 \tag{13}$$

et (12) devient

$$\begin{cases} \tilde{w}\nabla_y \tilde{u}^1 + \tilde{u}^1\nabla_y \tilde{w} + C\nabla_y p^1 = f^1 \\ \\ \nabla_y.\tilde{u}^1 = g^1 \end{cases} \tag{14}$$

avec

$$f^1 = -\nabla a^T\,[(u+w)_t + (u+w).\nabla(u+w) + \nabla(p+\pi) - \mu\nabla a^T\nabla a : \nabla_y\nabla_y w] \tag{15}$$

$$g^1 = -\nabla.w \tag{16}$$

Si on moyenne en y (14) multiplié respectivement par 1, $C^{-1}\tilde{w}$, $(\nabla a\nabla_y) \times w$ on obtient :

$$<f^1> + <g^1\tilde{w}> = 0 \tag{17}$$

$$<\tilde{w}^T C^{-1}f^1> + <(\tfrac{1}{2}\tilde{w}^T C^{-1}\tilde{w}+\pi)g^1> = 0 \tag{18}$$

$$<\nabla a^{-T}f^1.(\nabla a\nabla_y) \times (\nabla a^{-T}\tilde{w})> = 0 \tag{19}$$

$$<g^1> = 0 \tag{20}$$

(la dernière équation est déduite de (16)).

Enfin on remarque que (7) différentiée donne (14). Pour définir u^1 on doit donc rajouter les conditions

$$<\tilde{w}.C^{-1}\tilde{u}^1> = 0, \quad <w.(\nabla a \nabla_y) \times \nabla a^{-T}\tilde{u}^1> = 0, \quad <u^1> = 0 \tag{21}$$

(H.3) <u>Hypothèse</u> : Les conditions (17)-(18) sont les seules conditions de compatibilité de (14) et le système (14),(21) admet une solution isolée continuement différentiable par rapport C et r.

Ayant ainsi défini $\tilde{u}^1$, on calcule u^1 par (13) :

$$u^1 = \nabla a^{-T}(\tilde{u}^1+\sigma\tilde{w}+\rho\tilde{v}) \tag{22}$$

car le noyau de (14) contient $\tilde{w}$ et $\tilde{v} = \partial\tilde{w}/\partial r$. Les fonctions $\sigma(x,t)$, $\rho(x,t)$ seront déterminées ultérieurement.

5. EQUATIONS MOYENNEES.

Il est facile de montrer qu'avec (15) et (16), (17) devient :

$$u_t + u\nabla u + \nabla p + \nabla.<w \otimes w> = 0, \tag{23}$$

$$\nabla.u = 0, \tag{24}$$

que (18) devient

$$(\frac{\partial}{\partial t} + u.\nabla) <\frac{|w|^2}{2}> + <w \otimes w> : \nabla u + \nabla.<(\pi+ \frac{w^2}{2})w> + \mu<|(\nabla a\nabla_y)w|^2> \tag{25}$$

et que (19) devient

$$\begin{aligned}(\frac{\partial}{\partial t} + u\nabla)<w.(\nabla a\nabla_y) \times w> + <w \otimes (\nabla a\nabla_y) \times w> : \nabla u \\ - <(\mu\nabla a^T\nabla a : \nabla_y\nabla_y w.(\nabla a\nabla_y) \times w> + \nabla.<(\pi+\frac{1}{2}w^2).\nabla a\nabla_y \times w>\end{aligned} \tag{26}$$

La situation se présente donc de la façon suivante :

- H.2 permet de définir 2 <u>tenseurs</u> d'ordre 2

$$\underline{\underline{R}}(\nabla a^T \nabla a, \frac{r}{q}) = <\tilde{w} \otimes \tilde{w}> \; ; \quad \underline{\underline{S}} = <(\nabla a^{-T}\tilde{w}) \otimes (\nabla a \nabla)_y \times (\nabla a^{-T}\tilde{w}) \qquad (27)$$

fonction de ∇a et r (c'est-à-dire 10 paramètres en 3-D) ; 2 vecteurs

$$\underline{b}(\nabla a, \frac{r}{q}) = <(\pi + \frac{1}{2}\tilde{w}\, C^{-1}\tilde{w})\nabla a^{-T}\tilde{w}> \; ; \; c(\nabla a, \frac{r}{q}) = <(\pi + \frac{1}{2}\tilde{w}\, C^{-1}\tilde{w}).\nabla a \nabla_y \times (\nabla a^T \tilde{w})> \qquad (28)$$

et 2 scalaires :

$$\begin{cases} d(\nabla a, \frac{r}{q}) = <\nabla a^T \nabla a : \nabla_y \nabla_y w).(\nabla a \nabla_y) \times \nabla a^{-T} w> \\ e(\nabla a, \frac{r}{q}) = <|(\nabla a \nabla_y)(\nabla a^{-T} w)|^2> \end{cases} \qquad (29)$$

- l'écoulement moyen est solution du système couplé (9), (23)-(26) qui devient :

$$a_t + u\nabla a = 0 \; ; \; a(x,0) = x \qquad (30)$$

$$\begin{cases} u_t + u\nabla u + \nabla p + \nabla.(q\nabla a \underline{\underline{R}}(\nabla a^T \nabla a, \frac{r}{q})\nabla a^T) = 0 \\ \nabla.u = 0 \; ; \quad u(x,0) = u^o(x) \end{cases} \qquad (31)$$

$$\begin{cases} q_t + u\nabla q + q\nabla a \underline{\underline{R}}(\nabla a^T \nabla a, \frac{r}{q})\nabla a^T : \nabla u + \nabla.(b(\nabla a^T \nabla a, \frac{r}{q})q^{3/2}) + \mu e(\nabla a^T \nabla a, \frac{r}{q})q = 0 \\ q(x,0) = \frac{1}{2} <|w^o|^2> . \end{cases} \qquad (32)$$

$$\begin{cases} r_t + u\nabla r + q\underline{\underline{S}}(\nabla a, \frac{r}{q}) : \nabla u + \nabla.(c(\nabla a, \frac{r}{q})q^{3/2}) + \mu d(\nabla a, \frac{r}{q}) = 0 \\ r(x,0) = <w^o.\nabla \times w^o> \end{cases} \qquad (33)$$

Remarque. L'analogie avec le modèle phénoménologique de turbulence "k-ε" [7] est frappante ; l'énergie turbulente de dissipation est remplacée par l'hélicité r.

Malheureusement la dépendance de w par rapport à r (contrairement à la dépendance en q) n'est pas explicite.

6. CHAMPS INITIAUX w^o ISOTROPES, ECOULEMENTS MOYENS 2-D.

L'analyse montre que le terme en $\underline{\underline{R}}$ dans (31) n'est pas dissipatif. Par exemple, si w^o est aléatoire stationnaire isotrope, l'invariance du système (30)-(32) par rapport au changement de repère de référence implique [2] b=0 et :

$$\nabla a R(\nabla a^T \nabla a)\nabla a^T = \alpha I + \beta \nabla a \nabla a^T + \gamma(\nabla a \nabla a^T)^2$$

où α,β,γ sont des fonctions des invariants de $\nabla a^T \nabla a$. Pour des écoulements moyens bidimensionnels γ est nul et la positivité de $w \otimes w$ implique que β est positif. Donc, si $c,d \ll 1$, (30-(33) devient :

$$\left\{\begin{array}{l} a_t + u\nabla a = 0 \\ u_t + u\nabla u + \nabla.(q\beta(i)\nabla a\nabla a^T) + \nabla p = 0 \quad \nabla.u = 0 \\ q_t + u\nabla q + q[\beta(i)\nabla a\nabla a^T : \nabla u + \mu e(i)] = 0 \end{array}\right. \tag{34}$$

où i est la trace de $\nabla a \nabla a^T$.

Ce système a été simulé sur ordinateur dans le cas d'un écoulement dans une conduite (figure 1) et dans le cas d'un écoulement derrière un cylindre (figure 2).

7. CORRECTEURS.

Il est facile de voir que (34) linéarisé au voisinage de $\nabla a = I$ donne une équation des ondes pour a (voir [3] par exemple). Pour obtenir un terme de viscosité turbulente (bien connu des expérimentateurs) il faut poursuivre le développement asymptotique. Une analyse dimensionnelle confirme d'ailleurs que la viscosité turbulente (homogène à une vitesse x longueur = $w\varepsilon$) est de l'ordre de ε. Cette analyse sera présentée dans [6].

Pour calculer cette "viscosité turbulente" on pourrait continuer le développement asymptotique mais cela ne modifiera pas l'équation de u. En fait on souhaite inclure un terme en ε dans l'équation de u. Pour cela il suffit d'écrire les conditions de compatibilité sur l'équation de u^2 au lieu de u^1. Il faut donc remplacer les seconds membres de l'équation de u^1 par leurs projections sur l'espace des seconds membres admissibles et reporter la différence (nécessairement $O(\varepsilon)$) dans l'équation de u^2 ; c'est la technique du déport (voir [3]). Les conditions de compatibilité sur le système définissant u^2 donnent les équations du champ moyen (ces équations peuvent être obtenues formellement en moyennant directement (3) multipliée par 1, w, $(\nabla a^T \nabla) \times w$).

Les calculs font apparaître un terme supplémentaire dans l'équation de u :

$$-\varepsilon \nabla . (\underline{\underline{T}}(\nabla a) \nabla u)$$

qui, si $\underline{\underline{T}}$ est positif, défini, est bien un terme de viscosité. L'étude du signe de T est actuellement en cours sur ordinateur.

CONCLUSION.

On a montré comment l'homogénéisation permet d'obtenir des modèles de fermeture pour le champ moyen d'un écoulement "turbulent" au sens de (1). Cette étude repose sur l'hypothèse fondamentale (H.2) et ne sera donc d'un intérêt pratique que si cette hypothèse est vérifiée au moins numériquement et si les coefficients $\underline{\underline{R}}, \underline{\underline{S}}, \underline{b}$, d, e, peuvent être tabulés par ordinateur en fonction de ∇a ; il s'agit là d'un travail de longue haleine !

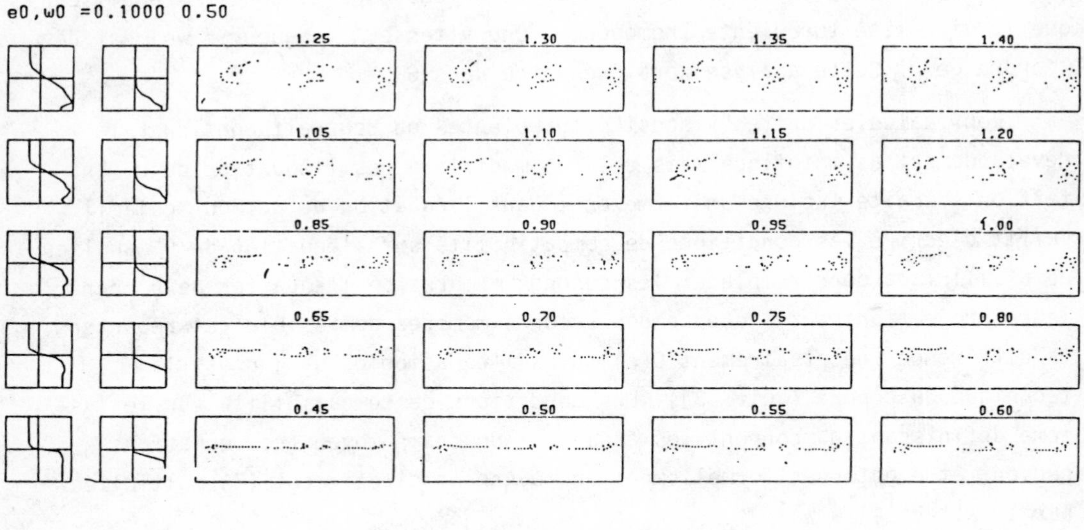

Figure 1(a)

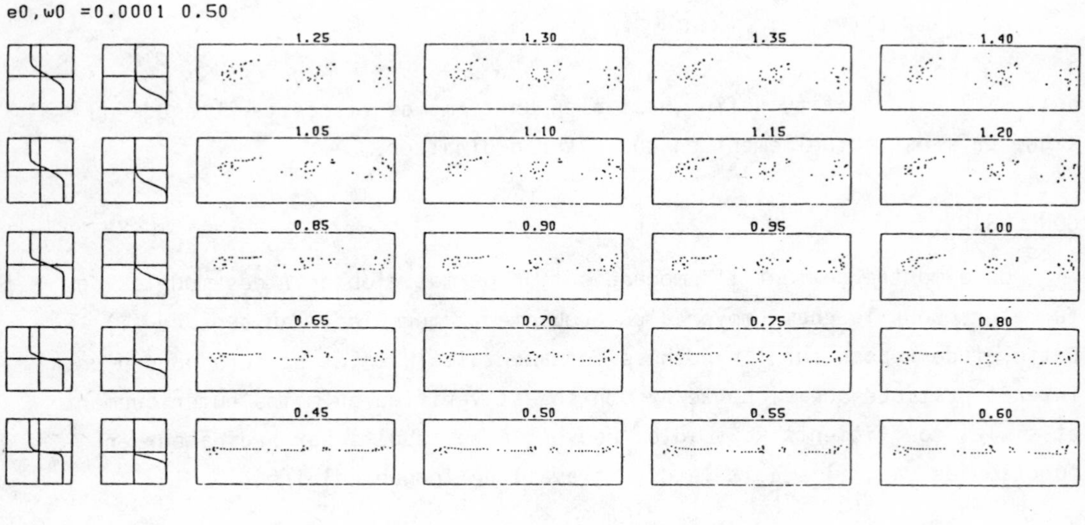

Figure 1(b)

Commentaire de la figure 1.

La figure 1 représente les résultats numérique d'un calcul de (34) sur ordinateur avec $q^o = 0.1$ (figure 1.a) et $q^o = 0$ (figure 1.b).

Ces résultats correspondent à la simulation numérique d'un jet plan : un jet de fluide arrive avec une vitesse uniforme dans un liquide au repos. Les cartouches rectangulaires représentent une portion du demi-jet supérieur aux instants dε temps 0.45, 0.50... L'interface fluide en mouvement fluide au repos, est simulé par une ligne de vortex comme le montre les cartouches ; cet interface se déforme et se bride conformément au phénomène dit de "Vortex sheet roll-up".

Les cartouches carrées représentent les moyennes horizontales de la vitesse horizontale (à gauche) et de l'énergie cinétique turbulente (à droite) pour les instants de temps 0.45, 0.65, 0.85, 1.05, 1.25.

En comparant les figures 1.a et 1.b, on voit que l'effet du modèle de turbulence (figure 1.a) est visible surtout sur les profils de vitesses horizontales moyennées (cartouches carrées gauches) : les profils de vitesse s'arrondissent, conformément à l'expérience.

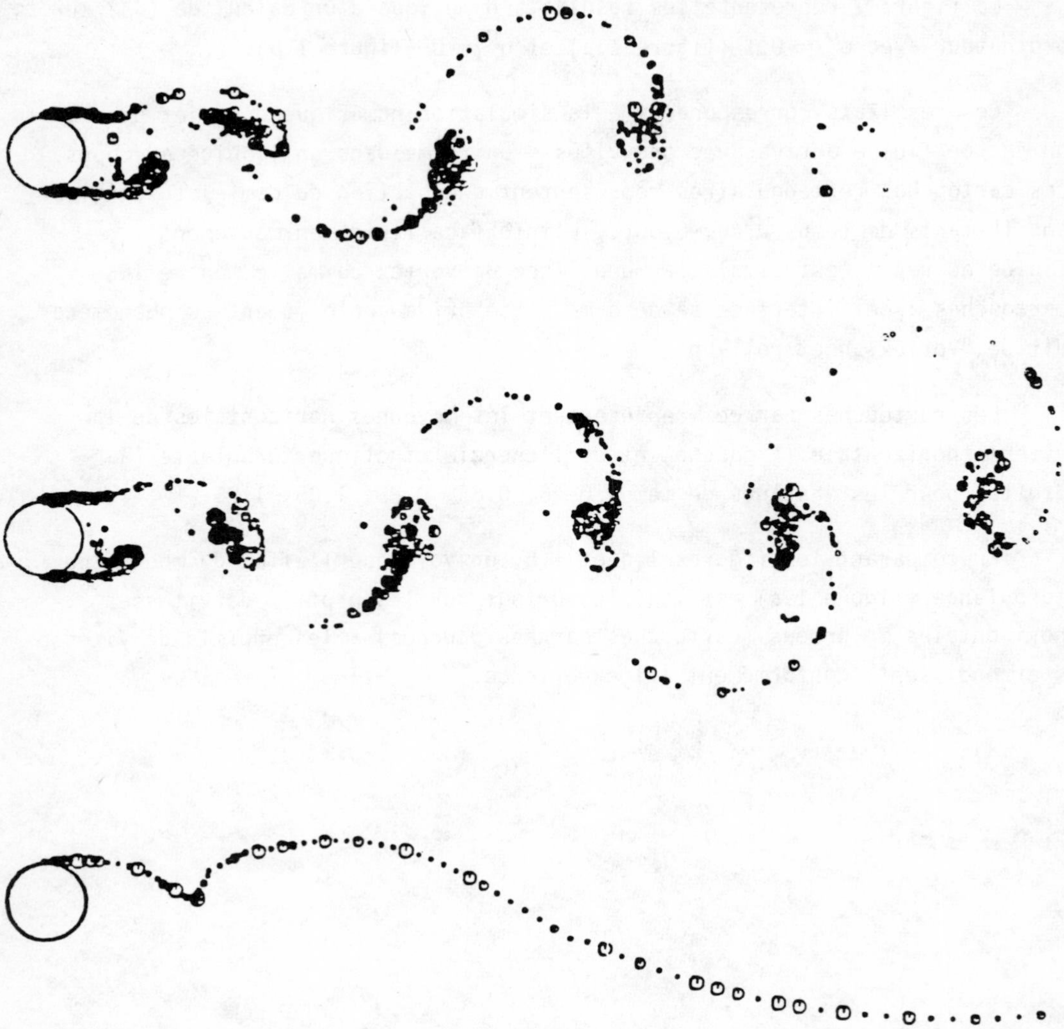

Figure 2 (from C. BEGUE, O. PIRONNEAU et al. [2])

Commentaire de la figure 2.

La figure 2 représente une simulation numérique d'écoulement turbulent plan derrière un cylindre, à l'aide de (34). On a représenté q(x,t), à t fixé à 2 valeurs (2 premières figures du haut). On a choisi des conditions aux limites de type Dirac en 2 points pour q, à la place de la condition initiale q^o, qui correspondent à la création d'énergie turbulente par la couche limite aux 2 points de décollement. Ainsi aux instants ultérieurs q est encore une somme de masses de Dirac dont les positions et intensités sont représentées par des petits cercles de diamètres proportionnels aux intensités.

La figure du bas représente l'histoire d'une masse de Dirac ; l'effet oscillant du terme $\langle w \times w \rangle$ est bien en évidence.

REFERENCES.

[1] A Pouquet, U. Frisch and J.P. Chollet, Turbulence with a spectral gap. Phys. Fluids 26 (4), (1983), 877-880.

[2] C. Begue, T. Chacon, D. Mc Laughlin, G. Papanicolaou, O. Pironneau, Convection of microstructures II. Proc. Numerical Methods in Engineering, Versailles 1983, R. Glowinski ed., North-Holland (to appear).

[3] D. Mc Laughlin, G. Papanicolaou, O. Pironneau, Simulation numérique de la turbulence par homogénéisation des structures de sous-mailles (to appear in SIAM).

[4] P. Perrier, O. Pironneau, Subgrid turbulence modelling by homogenization, Math. Modelling, Vol.2 (1981), 295-317.

[5] G. Papanicolaou, O. Pironneau, On the asymptotic behavior of motion in random flow. In Stochastic Nonlinear Systems, Arnold-Lefever, eds., Springer (1981).

[6] T. Chacon, O. Pironneau, On the mathematical foundation of the k-ε models (to appear).

[7] B.E. Launder, D.B. Spalding, Mathematical models of turbulence. Academic Press (1972).

O. PIRONNEAU
Université Paris-Nord et INRIA
(INRIA, 78153 LE CHESNEY)

Remerciements : Ce travail est le fruit d'une collaboration avec D.McLanghin et G. Papanicolaou.

C H TAUBES

A brief survey of the Yang–Mills–Higgs equations on $\mathbb{R}^3$

1. THE MONOPOLE PROBLEM.

The goal of this communication is to present some of the recent results and also unsolved questions for the Yang-Mills-Higgs equations on $\mathbb{R}^3$. To begin, a short introduction to this "monopole problem" is called for.

Yang-Mills-Higgs is a generic term which describes certain semi-linear, (degenerate) elliptic systems of differential equations on $\mathbb{R}^3$. A detailed introduction to these equations is provided in [1]. Here we restrict our attention to the so-called SU(2) theory. The differential equations are equations for twelve unknown functions. For economical reasons, these functions are grouped as a pair $c = (A,\Phi)$. At each $x \in \mathbb{R}^3$,

$$\Phi(x) = (\Phi^1(x),\Phi^2(x),\Phi^3(x))$$

represents three functions ; Φ is called the "Higgs field" in the physics literature. The A represents nine functions ; A is the "Yang-Mills potential" or "connection" :

$$A(x) = \begin{pmatrix} A_1^1 & A_1^2 & A_1^3 \\ A_2^1 & A_2^2 & A_2^3 \\ A_3^1 & A_3^2 & A_3^3 \end{pmatrix}$$

Thus, twelve unknowns.

Associated to each pair (A,Φ) is an energy or action functional. In order to write down this function, it is convenient to define first a "covariant derivative of Φ", nine functions

$$D_A\Phi = (D_A\Phi_j^i)_{i,j=1}^3 ,$$

where

$$D_A\Phi^1_j = \frac{\partial}{\partial x^j}\Phi^1 - A^2_j\Phi^3 + A^3_j\Phi^2,$$

$$D_A\Phi^2_j = \frac{\partial}{\partial x^j}\Phi^2 - A^3_j\Phi^1 + A^1_j\Phi^3,$$

$$D_A\Phi^3_j = \frac{\partial}{\partial x^j}\Phi^3 - A^1_j\Phi^2 + A^2_j\Phi^1.$$

Thus, $$D_A\Phi^i_j = \frac{\partial}{\partial x^j}\Phi^i + \text{quadratic }(A,\Phi).$$

The derivatives of A are most economically expressed through nine functions, together called the "Yang-Mills field strength" or "curvature", F_A. Here

$$F_A = (F^i_{Aj})^3_{i,j=1}$$

with

$$F^1_{A_1} = \frac{\partial}{\partial x^2}A^1_3 - \frac{\partial}{\partial x^3}A^1_2 - A^2_2A^3_3 + A^3_2A^2_3 \; ;$$

while the other F^i_{Aj} are obtained from this last expression by *cyclic* permutations of (1,2,3), i.e.,

$$F^2_{A_3} = \frac{\partial}{\partial x^1}A^2_2 - \frac{\partial}{\partial x^2}A^2_1 - A^3_1A^1_2 + A^1_1A^3_2.$$

Schematically, if one thinks of A as representing the three vectors $(A^i) = (A^i_j)$, then F_A represents the three vectors

$$\vec{F}^i_A = \text{curl }\vec{A}^i + \text{quadratic }(A),$$

where the symbol curl is the classical exterior derivative of vector calculus.

Given a pair (A,Φ) (12 functions) and given a real parameter $\lambda \in [0,\infty)$, one defines the λ-action of (A,Φ) to be

$$\mathcal{A}_\lambda(A,\Phi) = \frac{1}{2}\|F_A\|_2^2 + \frac{1}{2}\|D_A\Phi\|_2^2 + \frac{\lambda}{4}\|1 - |\Phi|^2\|_2^2 \qquad (1)$$

Here $\|.\|_2$ is the obvious L^2-norm on $\mathbb{R}^3$; i.e.

$$\|F_A\|_2^2 = \int_{\mathbb{R}^3} d^3x \sum_{i,j=1}^{3} |F^i_{Aj}|^2.$$

Also, $|\Phi|^2 = \sum_{i=1}^{3} (\Phi^i)^2$.

The $\lambda = 0$ case is rather special ; this is called the BPS limit (BPS is short for Bogomol'nyi [2]/Prasad-Sommerfield [3]).

The Yang-Mills-Higgs equations are the variational equations of $\mathcal{A}_\lambda$ for some $\lambda \in [0,\infty)$. These are also called the "monopole equations".

Définition. *A set of 12 fonctions* (A,Φ) *is a solution of the Yang-Mills-Higgs equations for some* λ *if*

1) (A,Φ) *are smooth.*
2) $\mathcal{A}_\lambda(A,\Phi) < \infty$
3) *For all compactly supported* (a,ϕ),

$$\frac{d}{dt}\mathcal{A}_\lambda(A+ta,\Phi+t\phi))\Big|_{t=0} = 0.$$

4) *For* $\lambda = 0$, *require that* $1 - |\Phi| \in L^6(\mathbb{R}^3)$.

The variational equations separate into the "Φ-equations" and the "A-equations". These Φ-equations are really three equations ;

$$\sum_{i=1}^{3} \left(\frac{\partial}{\partial x^i} D_A\Phi^1_i - A^2_i D_A\Phi^3_i + A^3_i D_A\Phi^2_i\right) + \lambda(1-|\Phi|^2)\Phi = 0$$

with the other two obtainable from this last by a cyclic permutation of (1.2.3). The A-equations are the three equations

$$\frac{\partial}{\partial x^2} F^1_{A3} - \frac{\partial}{\partial x^3} F^1_{A2} - A^2_2 F^3_{A3} + A^3_2 F^2_{A3} + A^2_3 F^3_{A2} - A^2_2 F^3_{A3} - \Phi^2 D_A\Phi^3_1 + \Phi^3 D_A\Phi^2_1 = 0$$

with the other two again obtainable through cyclic permutations of (1.2.3).

It is important to cut through the jungle of indices and new definitions. The crucial features of these equations are the following : First, $\mathcal{A}_\lambda$ is schematically

$$\mathcal{A}_\lambda = \frac{1}{2}\|\text{curl } \vec{A}\|_2^2 + \frac{1}{2}\|\nabla\Phi\|_2^2 + \text{nonlinear terms up to fourth order.}$$

Thus, the variational equations of $\mathcal{A}$ are schematically

$$\sum_{i=1}^{3} \frac{\partial^2}{\partial x^i \partial x^i} \Phi^j + \text{up to cubic nonlinearities} = 0.$$

$$\sum_{i=1}^{3} \left(\frac{\partial^2}{\partial x^i \partial x^i} A_k^j - \frac{\partial}{\partial x^i}\frac{\partial}{\partial x^k} A_i^j\right) + \text{up to cubic nonlinearities} = 0.$$

Observe that the Φ-equations are nonlinear Laplace equations ; while the A-equations are nonlinear generalizations of Maxwell's equations of magneto-statics.

The second crucial feature of the equations are that they have a degenerate-elliptic leading order symbol with semi-linear nonlinearities. These nonlinearities are subcritical for the Sobolev inequalities [4]. (In four dimensions these nonlinearities would be critical). Finally, the equations and the functionals $\mathcal{A}_\lambda$ are invariant under rotations and translations of $\mathbb{R}^3$. The translation invariance implies that this is a problem for which Palais-Smale Condition C fails (for a definition, see e.g. [5]).

The primary goal is : For a given $\lambda \geq 0$, find all solutions of the Yang-Mills-Higgs equations.

2. THE PHYSICS.

The action functional of Eq. (1) gives the energy of static solutions to a nonlinear wave equation on Minkowski space, $\mathbb{R}^3 \times$ time (cf. [6]). The critical points of this action functional (the finite action solutions to the Yang-Mills-Higgs equations) are *solutions* for this nonlinear wave equation. (The nonlinear wave equation in question is defined by the *Lagrangian* below, $\mathfrak{L}$, which is constructed from time dependent (A,Φ) (twelve functions on $\mathbb{R}^3 \times$ time) :

$$\mathfrak{L}(A,\Phi) = \tfrac{1}{2}\{ \left|\frac{\partial\Phi}{\partial t}\right|^2 + \left|\frac{\partial A}{\partial t}\right|^2 - |D_A\Phi|^2 - |F_A|^2 - \tfrac{\lambda}{2}(1-|\Phi|^2)^2\}).$$

The wave equation itself is a simplified model of equations which physicists believe describe the forces between subnuclear particles. Today, physicists believe that there are four different forces in nature : (1) gravity (which holds us to the earth, and holds the earth in orbit around the sun) ; (2) electro-magnetism (all other macroscopic forces are electro-magnetic ; it is responsible for electricity and magnetism) ; (3) the weak force (many forms of radioactivity are due to the weak force) ; (4) the strong force (the release of nuclear energy in power plants or bombs).

Yang-Mills-Higgs equations are supposed to provide the mathematical basis for theories which explain the electro-magnetic, weak, and strong forces as the low energy facets of a single unified force (much as Maxwell's equations describe the unification of electricity and magnetism). These unifying theories are named GUTS, for Grant Unified Theories [7]. For example, the SU(2) model under consideration here is a "baby" version which describes a unification of the weak force with electro-magnetism. The physicist interprets the field (A,Φ) in the following way : The field

$$B_i(\vec{x}) = \sum_{j=1}^{3} \Phi^j F_A^j(x) \tag{2}$$

is interpreted as the usual magnetic field. If one allowed time dependent (A,Φ), then

$$E_i(x,t) = \sum_{j=1}^{3} \Phi^j \frac{\partial A_i^j}{\partial t}$$

would be the usual electric field. Meanwhile,

$$A_i^j - \Phi^j\left(\sum_{j=1}^{3} \Phi^k A_i^k\right)$$

represents a *vector potential* for the weak force.

The equations of interest here are now known to admit soliton solutions. If these equations actually *model* physical reality, then the solitons should be the mathematical description of some real physical entity. The physicists believe that the solitons represent mathematically a new fundamental particle called a magnetic monopole. That is, the magnetic field in Eq. (2) of this soliton has a large distance multi-pole expansion given by

$$B_i(x) = n \;.\; \frac{x_i}{|x|^3} + \sum_{j=1}^{3} (\ell_j x_j x_i - \tfrac{1}{3}|x|^2 \ell_i)\, \frac{1}{|x|^5} + O(|x|^{-4}).$$

Here, n is the monopole charge and $\{\ell_i\}$ is the dipole moment. The remainder contains the quadropole and higher order terms.

It is a fundamental prediction of Maxwell's equations that *no* magnetic field can ever have a monopole term - Maxwell says : all magnets are dipoles. Test this with the magnets you buy in the store - they will all have two poles. All magnets found to date have two poles, called north and south ; and when you cut them in half, *each half* has both north and south too :

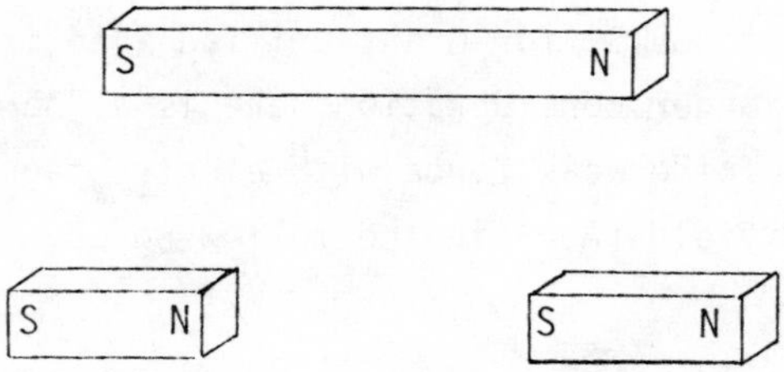

The mathematics of this unified theory makes a *fundamentally new* prediction : there exists objects (the solitons) which have only one magnetic pole

Ⓝ or Ⓢ ;

with just a north (called the "monopole") or just a south (called the "anti-monopole").

If Grand Unified Theories correctly describe nature, then magnetic monopoles exist. Presently, experimenters are looking for the predicted monopoles [8]. In fact, the detector is simple enough (in principle) to build yourself at home. The discoverer of a magnetic monopole will surely win a Nobel prize, so here's your big chance. Take a battery and a light bulb and a coil of wire and construct the following circuit :

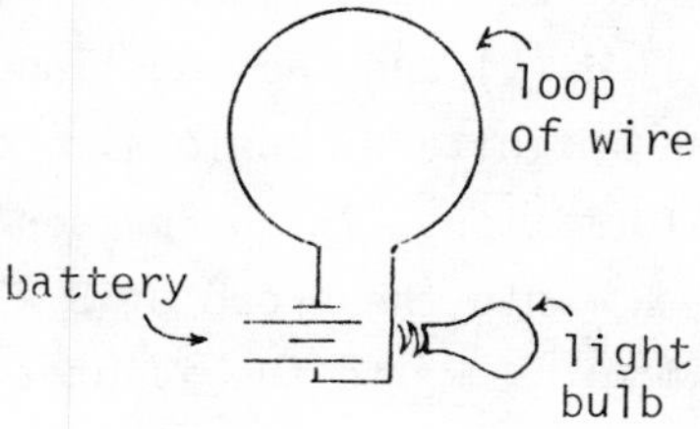

Should a monopole pass through the loop of wire, the light will *increase* momentarily in brightness. Should an anti-monopole pass through (or a monopole in the other direction), the light will decrease momentarily in brightness. A dipole going through will cause the light to first increase and then decrease (or the reverse).

3. THE MATHEMATICAL ISSUES.

The Yang-Mills-Higgs equations come from a variational problem which is (1) (degenerate) elliptic, (2) semi-linear and subcritical, and (3) translationally invariant. These three properties are the crucial ones. The difficulties which arise here also arise in your own favorite variational problem with properties (1)-(3). Don't let the tangle of indices and unknown functions confuse you ; it is (1)-(3) above that determine the phenomena which occur.

The problem posed is to find the solutions. To find them by a variational argument means invariably a discussion of the *convergence* of a sequence of configurations (for example, a minimizing sequence ; or min-max). As in classical variational problems, the convergence question has two aspects. The first question concerns convergence up to the boundary - or here - global convergence on $\mathbb{R}^3$.

If your favorite problem satisfies Condition C on an appropriate Banach space, then the first question is answered in what is now a standard way. The Sobolev inequalities (subcritical nonlinearities) determine when Condition C is satisfied on bounded domains (see, e.g., [9]).

For Yang-Mills-Higgs, the problem is that the equations are degenerate elliptic. Check that the leading order symbol for the "A" equation is positive definite, only semi-definite. (Recall that $\mathcal{A}_\lambda(A,\Phi) = \frac{1}{2}\|\text{curl } \vec{A}^i\|_2^2 + ..$). The reason is that the functional $\mathcal{A}_\lambda(.)$ and the equations themselves are invariant under the action on an infinite dimensional group of transformations called the group of gauge transformations. In the present context, a gauge transformation is a C^∞ map from $\mathbb{R}^3$ into the group of 3×3 special orthogonal matrices (the group (SO(3)) taking $0 \in \mathbb{R}^3$ to the identity matrix. If $\Lambda^{ij}(x)$ $(i,j=1,2,3)$ is such a gauge transformation ; and if (A,Φ) is a configuration, then $(\Lambda A,\Lambda\Phi)$ is the configuration with

$$(\Lambda A)^1_j = \sum_{k=1}^{3} [\Lambda^{1k} A^k_j + \frac{1}{2} \Lambda^{2k} \frac{\partial}{\partial x^j} \Lambda^{3k} - \frac{1}{2} \Lambda^{3k} \frac{\partial}{\partial x^j} \Lambda^{2k}]$$

and $(\Lambda A)^2_j$, $(\Lambda A)^3_j$ is obtained from this last equation by cyclic permutations of (1,2,3). Also, for each $i \in (1,2,3)$,

$$(\Lambda\Phi)^i = \sum_{k=1}^{3} \Lambda^{ik} \Phi^k .$$

Convince yourself that $\mathcal{A}_\lambda((\Lambda A,\Lambda\Phi)) = \mathcal{A}_\lambda(A,\Phi)$ and that the Yang-Mills-Higgs equations are also invariant: A historical note here is that these equations were *designed* to have this invariance.

Thanks to Karen Uhlenbeck's work, this invariance poses no major problem with respect to the question of local convergence. She has provided the tools to handle the question. In dimensions less than 4, Yang-Mills-Higgs satisfies a modified, gauge-invariant Condition C on bounded domains ; 4 is the critical dimension. Required reading is Uhlenbeck's "Connections with L^p bounds on curvature" [10] in which the gauge invariant Sobolev inequalities are derived. (See also [11,§5] for the generalization to Yang-Mills-Higgs, and [12],[13] for applications in dimension 4. See [14] for a discussion of the global gauge fixing problem).

The question of convergence for min-max sequences globally on $\mathbb{R}^3$ is the second convergence question. Here, one must analyze whether or not information "drifts off to $|x| = \infty$". The author's new techniques in [15] allow one to analyze whether global convergence on $\mathbb{R}^3$ occurs. The essential idea is to obtain a priori estimates on min-max cells which are strong enough to allow a "long range force" analysis. If solitons at large distances "attract" each other, then convergence can be proved. No more will be said here, the reader is referred to §1,2 of [15].

4. A BRIEF SURVEY OF PUBLISHED RESULTS.

Before beginning, it is important to remark on the fact that $\mathcal{A}_\lambda$ is not finite on the function space $\times_{12} C^\infty(\mathbb{R}^3)$. The usefult configuration space is

$$\mathfrak{C}^\lambda = \{(A,\Phi) \in \times_{12} C^\infty(\mathbb{R}^3) : \mathcal{A}_\lambda(A,\Phi) < \infty \text{ and if } \lambda = 0, \; 1 - |\Phi| \in L^6(\mathbb{R}^3)\}.$$

This space has a natural topology for which it deforms by homotopy onto the space of smooth maps from S^2 to S^2, a space whose homotopy and homology groups are understood [16]. An important fact is that each $\mathfrak{C}^\lambda$ has a countable set of path components :

Theorem 1. (Groisser [17]) : *For* $\lambda \in [0,\infty)$, *each* $\mathfrak{C}^\lambda$ *is the disjoint union over* $n \in \mathbb{Z}$ *of path components* $\{\mathfrak{C}_n\}$. *for* $(A,\Phi) \in \mathfrak{C}_n^\lambda$,

$$(4\pi)^{-1} \langle F_A, D_A\Phi \rangle_2 = n.$$

Physicists interpret $\mathfrak{C}_n^\lambda$ as being the set of configuration which have monopole charge equal to n.

An immediate consequence of theorem 1, observed first by Bogomol'nyi [2], is that

$$\mathcal{A}_\lambda|_{\mathfrak{C}_n^\lambda} \geq 4\pi|n|.$$

And, equality holds if and only if $\lambda = 0$ with

$$F^i_{Aj} = \text{sign}(n) D_A \Phi^i_j . \tag{3}$$

This last equation is called the Bogomol'nyi equation. For $n = 0$, the Bogomol'nyi equation has only the trivial solution ($A = 0$, $\Phi = (1,0,0)$) and its gauge transforms. For $n = \pm 1$, an explicit, spherically symmetric solution to Eq. (3) was discovered by Prasad and Sommerfield [3]. It is unique up to gauge transformation and translations on $\mathbb{R}^3$. For $|n| > 1$, the Bogomol'nyi equations are well understood (see [18] for a review). The initial results were

Theorem 2. (C. Taubes, see [19]) : *For each* $n \in \mathbb{Z}$, $\mathcal{A}_{\lambda=0}$ *achieves its minimum on* $\mathcal{C}^0_n$. *Up to gauge equivalence, there exists a* $4|n|$ *parameter family of solutions to the Bogomol'nyi equations.*

The result in theorem 2 was an existence theorem. Subsequently, R. Ward 20 showed how Eq. (3) can be interpreted as the integrability condition for a complex structure on a vector bundle over the complex 2-manifold $T\mathbb{C}\mathbb{P}^1$ (the tangent bundle to S^2). This opened the way for a complex algebraic analysis of Eq. (3) which culminated in the following remarkable result of S.K. Donaldson :

Theorem 3. (S.K. Donaldson [21]) : *The solutions to Eq.* (3) *in* $\mathcal{C}^0_n$ *are, up to gauge transformations, naturally equivalent to the space of degree homomorphic maps from* $\mathbb{CP}^1$ *to* $\mathbb{C}\mathbb{P}^1$ (S^2 *to* S^2) *which take the north pole to itself.*

The existence proof of theorem 2 does not use variational arguments. However, variational techniques have been the major tool for studying the non-minimal critical points of $\mathcal{A}_\lambda$. The most complete results are for $\lambda = 0$.

Theorem 4. (C. Taubes [11]) : *For* $\lambda = 0$, *there exists a non-minimal critical point of* $\mathcal{A}_0$, *in* $\mathcal{C}_0$, *which is obtainable by a min-max procedure over non-contractible loops in* $\mathcal{C}_0$.

The result above suggests that $\mathcal{A}_0$ might truly have all of its "topologically required" (see [16]) critical points. What we mean here is the following : Let $\mathcal{F}$ be a family of compact subsets of $\mathcal{C}^\lambda$ with the property that for each (continuous) homotopy $\phi : [0,1] \times \mathcal{C}^\lambda \to \mathcal{C}^\lambda$ ((0,.) = identity), the assertion $F \in \mathcal{F}$ implies that $\phi(1,F) \in \mathcal{F}$. Such a family $\mathcal{F}$ is called a "homotopy invariant family of compact subsets of $\mathcal{C}^\lambda$". To each such $\mathcal{F}$, one associates the number

$$\mathcal{A}_{\mathcal{F}}^\lambda = \inf_{F \in \mathcal{F}} \sup_{c \in F} \mathcal{A}_\lambda(c).$$

Each $\mathcal{A}_{\mathcal{F}}^\lambda$ is a potential critical value of $\mathcal{A}_\lambda$. Were $\mathcal{C}^\lambda$ a compact manifold with $\mathcal{A}_\lambda$ a smooth function, then Morse theory would insure that each $\mathcal{A}_{\mathcal{F}}^\lambda$ was a critical value. The associated critical point would be "topologically required".

Theorem 5. (C. Taubes [15]) : *For* $\lambda = 0$ *let* $n \in \mathbb{Z}$. *Let* $\mathcal{F}$ *be a homotopy invariant family of compact subsets of* $\mathcal{C}_n^0$. *There is a critical point of* $\mathcal{A}_0$ *in* $\mathcal{C}_n^0$ *with critical value* $\mathcal{A}_{\mathcal{F}}^0$.

Using theorem 5, and the fact that for each $n \in \mathbb{Z}$ and $\lambda \geq 0$, the set

$$\text{crit}_n^\lambda = \{\mathcal{A}_{\mathcal{F}}^\lambda : \mathcal{F} \text{ is a homotopy invariant family of compact subsets of } \mathcal{C}_n^\lambda\}$$

is unbounded [16], and one obtains

Theorem 6. (C. Taubes [15]) : *For* $\lambda = 0$, *and for each* $n \in \mathbb{Z}$, $\mathcal{A}_0$ *has an unbounded set of critical values.*

For positive λ, the functionals $\mathcal{A}_\lambda$ are not well understood. However there are some results ; all use min-max. First, there is always the trivial solution $(A=0, \Phi=(1,0,0)) \in \mathcal{C}_0^\lambda$ and its gauge transforms. This has action zero ; it is the global minimum of $\mathcal{A}_\lambda$. For every $\lambda > 0$, spherically symmetric solutions are known to exist in $\mathcal{C}_{\pm 1}^{\lambda}$ [22],[23]. Also, it is known that for λ sufficiently small, $\mathcal{A}_{\lambda}$ achieves its infimum on $\mathcal{C}_{\pm 1}^{\lambda}$ (D. Groisser [24]). An unpublished argument of that author implies that for λ small, these spherically symmetric solutions are Groisser's minima.

For non-minimal critical points of $\mathcal{A}_{\lambda}$, $\lambda > 0$, min-max techniques yield

Theorem 7. (D. Groisser [24]) : *For $\lambda > 0$, but sufficiently small, there exists a non-minimal critical points of $\mathcal{A}_{\lambda}$ in $\mathcal{C}_{0}^{\lambda}$.*

It is a near sure bet that Groisser's techniques can establish theorem 7 for all $\lambda \geq 0$.

Do other critical points exist for $\mathcal{A}_{\lambda}$ when $\lambda > 0$? The following is, perhaps, a reasonable conjecture :

Conjecture. *For $\lambda > 0$, it is only for $n = 0, \pm 1$ that $\mathcal{A}_{\lambda}$ on $\mathcal{C}_{n}^{\lambda}$ achieves its infimum.*

Conversations with J.M. Coron and A. Bahri have convinced this author to conjecture that, at least for λ small, there are (possibly) non-minimal critical points of $\mathcal{A}_{\lambda}$ in $\mathcal{C}_{n}^{\lambda}$ for $n \neq 0$. Here is the reasoning (with the help of Mssrs. Coron and Bahri) : For $\lambda = 0$, the $4|n|$ dimensional manifold of minima in $\mathcal{C}_{n}^{0}$ (the set of gauge equivalence classes of solutions to Eq. (3) in $\mathcal{C}_{n}^{0}$) is topologically nontrivial (see [21],[25] and [15]). Presumably, it does not retract onto its boundary. For λ small, there is a way to continuously embed this manifold into $\mathcal{C}_{n}^{\lambda}$ with action $\mathcal{A}_{\lambda}$ uniformly near the infimum of $\mathcal{A}_{\lambda}$ on $\mathcal{C}_{n}^{\lambda}$. Min-max over homotopies of this embedding may yield either a minimum for $\mathcal{A}_{\lambda}$ on $\mathcal{C}_{n}^{\lambda}$, or else a non-minimal critical point for $\mathcal{A}_{\lambda}$ on $\mathcal{C}_{n}^{\lambda}$ with critical value near the infimum. To follow the argument, consider how min-max might prove the existence of the unique (non-minima) critical point of the function $f(x) = \exp(-|x|^2)$ on $\mathbb{R}^n$. There is no compactness here, but using the fact that $\mathbb{R}^n$ does not retract onto S^{n-1}, one can readily give a min-max existence proof.

As a final remark, the reader should be well aware of the fact that this SU(2) model is only the simplest of the Yang-Mills-Higgs equations. For any compact Lie group, G, there are the analogous functionals to $\mathcal{A}_{\lambda}$ (see [1]). For $\lambda = 0$, the existence question for minima is now reasonably well understood (see [19],[26]). The existence or non-existence of non-minimal critical points is a question which has yet to be studied.

REFERENCES.

[1] A. Jaffe and C.H. Taubes, *Vortices and Monopoles*, Birkhauser, Boston, 1980.

[2] E.B. Bogomol'nyi, The stability of classical solutions . Sov. J. Nucl. Phys. 24, 449 (1976).

[3] M.K. Prasad and C. Sommerfield, Exact classical solutions for the 't Hooft monopole and the Julia-Zee dyon . Phys. Rev. Lett. 35, 760 (1975).

[4] K.K. Uhlenbeck, Variational problems for gauge fields . in *Seminar on Differential Geometry*, S.T. Yau, ed., Princeton University Press, Princeton, New Jersey, 1982.

[5] R. Palais, Critical point theory and mini-max principle . Proc. Symp. Pure Math., Vol. 15, American Math Society, Providence, Rhode Island (1970).

[6] D. Eardly and V. Moncrief, The global existence of Yang-Mills-Higgs fields in 4-dimensional Minkowski space . Part I : Commun. Math. Phys. 83, 171 (1982). Part II : Commun. Math. Phys. 83 (1982).

[7] G. 't Hooft, Gauge theories of the forces between elementary particles . Sci. Amer. 242, n°6, pp. 104-138, (June 1980).

[8] R. Carrigan, Jr. and W.P. Trower, Superheavy magnetic monopoles . Sci. Amer. 246, n°4, pp. 106-118, (April 1982).

[9] M. Berger, *Nonlinearity and Functional Analysis*. Academic Press, New York, 1977.

[10] K.K. Uhlenbeck, Connections with L^p-bounds on curvature . Commun. Math. Phys. 83, 31 (1981).

[11] C.H. Taubes, The existence of a nonminimal solution to the SU(2) Yang-Mills-Higgs equations on $\mathbb{R}^3$. Part I : Commun. Math. Phys. 86, 257 (1982). Part II : Commun. Math. Phys. 86, 299 (1982).

[12] S. Sedlacek, A direct method for minimizing the Yang-Mills functional over 4-manifolds . Commun. Math. Phys. 86, 515 (1982).

[13] C.H. Taubes, Path connected Yang-Mills moduli spaces . Jour. Diff. Geom., to appear.

[14] R. Schoen and K.K. Uhlenbeck, Regularity of minimizing harmonic maps into the sphere . Inven. Math., to appear.

[15] C.H. Taubes, Min-max for the Yang-Mills-Higgs equations . Commun. Math. Phys., to appear.

[16] C.H. Taubes, Monopoles and maps from S^2 to S^2 ; the topology of the configuration space . Commun. Math. Phys., to appear.

[17] D. Groisser, Integrality of the monopole number in SU(2) Yang-Mills-Higgs theories on $\mathbb{R}^3$. Commun. Math. Phys. 93, 367 (1984).

[18] J. Borzlaff, Magnetic poles in gauge field theories . Communications of the Dublin Institute for Advanced Studies, A 27 (1983).

[19] C.H. Taubes, Existence of multi-monopole solutions to the non-abelian Yang-Mills-Higgs equations . Commun. Math. Phys. 80, 343 (1981).

[20] R. Ward, A Yang-Mills monopole of charge 2 . Commun. Math. Phys. 79, 317 (1981).

[21] S.K. Donaldson, Nahm's equations and the classification of monopoles . Commun. Math. Phys., to appear.

[22] G. 't Hooft, Magnetic monopoles in unified gauge theories . Nucl. Phys. B79, 276 (1974) ; also A.M. Polyakov, Particle spectrum in quantum field theory . JETP Lett. 20 194 (1974).

[23] R. Weder, Existence, regularity and exponential decay of finite energy solutions to gauge field equations . Communicaciones Técnicas 10, 198 (1979).

[24] D. Groisser, SU(2) Yang-Mills-Higgs theory on $\mathbb{R}^3$. Harvard University Ph.D. Thesis, 1983.

[25] G. Segal, Topology of spaces of rational functions . Acta Math. 143, 39 (1979).

[26] M.K. Murray, Monopoles and spectral curves for arbitrary Lie groups. Commun. Math. Phys. 90, 263, (1983).

Clifford H. TAUBES
Department of Mathematics
University of California
Berkeley
CALIFORNIA 94720
USA

R TEMAM

Attractors for Navier–Stokes equations

1. INTRODUCTION.

The aim of this lecture is to present the latest results concerning the attractors for the Navier Stokes equations. The study of the Navier-Stokes equations (N.S.E.) as an infinite dimensional dynamical system is motivated by the understanding of turbulence and chaos in fluids. When a fluid is driven by a sufficiently strong external excitation, the permanent flow which appears is not laminar and is not stationary even if the exciting forces are time independant. The natural mathematical object which represents such a flow is then a subset of the appropriate function space which enjoys important properties and in particular that of being invariant for the semi-group associated to the Navier-Stokes evolution equation. This set is called a functional invariant set and may or may not be attracting in the function space, although sometimes, by an abuse of language, the term attractor is used instead of functional invariant set. The study of these sets can (hopefully) help understand and describe turbulent flows.

The results which are presented here are the most precise one available at the moment and are borrowed from A.V. Babin - M.I. Vishik [2], P. Constantin - C. Foias - R. Temam [5], P. Constantin - C. Foias - O. Mauley - R.T. [6], R.T. [30]. The first results on the attractors for the N.S.E. were derived in [11] ; other results are mentioned below or in the previous references. Since the initial value problem for the N.S.E. is well set in dimension 2 and is not known to be well set in dimension 3 (cf. J. Leray [19],[20],[21], J.L. Lions [23]), the results which we present are different in both cases. Sec. 2 is devoted is the two-dimensional case and Sec. 3 to the three dimensional case. More precise results are obtained in dimension 2 but, in dimension 3, an interesting connection is made with the Kolmogorov approach to turbulence and we show how one can prove rigorously an heuristic estimate by Landau - Lifschitz [18] on the number of degrees of freedom of a turbulent flow. Sec. 4 contains some ideas of the proofs of the main results of Sec. 2 and 3 and Sec. 5 gives some indications on other problems : non homogeneous boundary condition, Bénard problem. Let us mention also that some comments on the relevance of the results described here to computational fluid dynamics are given in Sec. 4.

In the rest of this introduction we briefly recall the Navier-Stokes equations and their mathematical setting. □

The Navier Stokes equations for an incompressible viscous fluid with density 1 are written

$$\frac{\partial u}{\partial t} - \nu\Delta u + (u.\nabla)u + \nabla p = f \tag{1.1}$$

$$\nabla.u = 0 \tag{1.2}$$

where $u = (u_1,u_2)$ or (u_1,u_2,u_3) is the velocity vector, p is the pressure ; $\nu > 0$ is the kinematic viscosity and f represents the density of volumic force (the driving force) per unit volume. The results presented in Sec. 2 and 3 apply to two cases : that of the flow in a bounded domain Ω of $\mathbb{R}^n$ (n=2 or 3), with a rigid Γ at rest (u=0 on Γ, $\forall t$) or the flow in $\mathbb{R}^n$ with a space periodicity boundary condition, u and p being periodic of period L_i in each direction x_i, i=1,2 or i=1,2,3 ; in this case we denote by Ω the period cell, $\Omega = \prod_{i=1}^{n} (0,L_i)$.

In the functional setting of the N.S.E. (Eqs. (1.1)(1.2) completed with the boundary and initial conditions), we consider u as a function of t with values in a Hilbert space H, which is an appropriate Hilbert subspace of $L^r(\Omega)^n$. Then the N.S.E. become an infinite dimensional dynamical system, corresponding to the differential equation in H :

$$\frac{du}{dt} + \nu Au + B(u,u) = f, \quad t > 0 \tag{1.3}$$

$$u(0) = u_o. \tag{1.4}$$

Here A is a linear unbounded positive self-adjoint opeator in H with domain $D(A) \subset H$, and B(.,.) is a bilinear compact operator from $D(A) \times D(A)$

into H which enjoys many other continuity properties ([2]) ; the reader is referred to R.T. [29] for the details on the functional setting of the Navier-Stokes equations.

2. THE TWO DIMENSIONAL CASE.

We assume that f is given independant of time, so that the dynamical system (1.3)(1.4) is autonomous :

$$f(t) \equiv f \in H \tag{2.1}$$

It is known [23][29] that in dimension 2 the initial value problem (1.3)(1.4) is well set for every $u_o \in H$, the unique solution of (1.3)(1.4) being furthermore analytic in time with values in D(A). This allows us to define a family of operators

$$S(t) : u_o \to u(t), \quad t > 0,$$

which are continuous from H into D(A) (S(0) = I is continuous from H into H), and enjoy the usual semi-group properties

$$S(t).S(s) = S(t+s), \; s,t \geq 0.$$

We now define a functional invariant set for (1.3) :

Definition. *Functional Invariant Set.*

A functional invariant set for (1.3) *(or for* S(t)) *is a set* $X \subset H$ *such that*

$$S(t)X = X, \quad \forall t > 0. \tag{2.2}$$

([2]) The powers of A, A^α, $\alpha \in \mathbb{R}$ are well defined linear self-adjoint operators with domains $D(A^\alpha)$; B is continuous from $D(A^{\alpha_1}) \times D(A^{\alpha_2})$ into $D(A^{\alpha_3})$, the relations between α_1, α_2, α_3, depending on the space dimension n.

Since $S(t)$ maps H into $D(A)$, such a set X is necessarily included in $D(A)$.

Such a set X is said to be an *attractor* (in H, or $D(A^\alpha)$ for some $\alpha \in \mathbb{R}$), if there exists a neighborhood $\mathcal{O}$ of X (in H or $D(A^\alpha)$) such that, for every $u_0 \in \mathcal{O}$, $S(t)\, u_0$ converges to X (in H or $D(A^\alpha)$), as $t \to +\infty$.

It plainly follows from (2.2) that if $u_0 \in X$, then $S(t)\, u_0 \in X$, $\forall t > 0$; but it also follows from (2.2) that for every $u_0 \in X$ the backward initial value problem (1.3)(1.4) has a solution for every $t < 0$ and $u(t) \in X \quad \forall t < 0$. The set X can be reduced to a stationary (time independant), solution u_s of (1.3), or to the trajectory

$$\{u(t), \quad t \in [0,T]\}$$

of a time periodic solution of (1.3) ($u(T) = u(0)$ for some $T > 0$) or to a possibly more complicated set.

Given $u_0 \in H$ (and $f \in H$), we consider the set

$$X = X(f,u_0) = \bigcap_{s>0} \overline{\bigcup_{t\geq s} \{S(t)u_0\}} \tag{2.3}$$

where the closures are taken in H. It was shown in [11] that $X(f,u_0)$ is a functional invariant set for (1.3) and that the solution $u(.)$ of (1.3)(1.4) converges, in H, to $X(f,u_0)$ as $t \to \infty$. *Thus* $X(f,u_0)$ *is the subset of* H *which describes the permanent flow* (i.e. *the flow after a transient period) for* f *and* u_0 *given.*

In order to introduce a larger functional invariant set (the universal attractor) we first define the absorbing set

Definition. *Absorbing Set.*

Let $V = D(A^{1/2})$ ($D(A) \subset V \subset H$). We define with V. Arnold [1] (see also M.I. Vishik [32] in this volume), an absorbing set in V for (1.3) : *a set* $\mathcal{A} \subset V$, *bounded in* V *is absorbing in* V *if for every bounded set* $\mathcal{B} \subset V$, *there exists* $t_0 = t_0(\mathcal{B})$ *such that*

$$S(t)\mathcal{B} \subset \mathcal{A}, \textit{ for } \quad t \geq t_0(\mathcal{B}) \tag{2.4}$$

We have

Lemma 2.1. *There exists a set* $\mathcal{A}$ *bounded in* V *which is absorbing for* (1.3).

The proof of this lemma is an extension of that of C. Foias - G. Prodi [10] who prove that every particular solution of (1.3)(1.4) with $u_0 \in V$ is uniformly bounded in V for $t \geq 0$ (see also [11]).

We now consider

$$\tilde{X} = \tilde{X}(f) = \bigcap_{s>0} \overline{\bigcup_{t \geq s} \{S(t)\mathcal{A}\}} \tag{2.5}$$

This set is a functional invariant set ; it is an attractor in H or V since every solution of (1.3)(1.4) converges to it as $t \to +\infty$. It contains all the sets $X(f,u_0)$, it is the largest functional invariant set (or attractor) bounded in H or V : we call it the *universal attractor* [11] ([3]). The universal attractor enjoys the following properties :

Theorem 2.1. (regularity)

If Γ *is a* C^∞ *manifold* ([4]) *and* $f \in C^\infty(\bar{\Omega})^2 \cap H$, *then* $\tilde{X}(f) \subset C^\infty(\bar{\Omega})^2$.

Theorem 2.2. (finite dimensionality)

The Hausdorff and fractal dimensions of the universal attractor $\tilde{X}(f)$ *are finite* ($\forall f \in H$).

Of course theorems 2.1 and 2.2 apply as well to any functional invariant set X bounded in H (or V). Theorem 2.1 is proved in C. Guillopé [17]. The result concerning the Hausdorff dimension in theorem 2.2 is proved in [11] but the proof given there extends easily to the fractal dimension as well. For the definitions of the Hausdorff and fractal dimensions see H. Federer [7], B. Mandelbrot [24].

([3]) It is called the maximal attractor in M.I. Vishik [32]

([4]) This assumption in the case of the boundary condition u=0 on Γ ; no assumption in the space periodic case.

Physical bound on the dimension.

Theorem 2.2 shows that a two dimensional flow depends on a finite number of parameters. It is then interesting to try to estimate this dimension in term of the data and, for that purpose, we introduce a non-dimensional number

$$G = \frac{|f|\ L_0^2}{\nu^2}$$

where $|f|$ is the norm of f in H, ν is the kinematic viscosity and L_0 is a typical length of Ω, its diameter for example. Some authors (see [2][14]) prefer to introduce a Reynolds number

$$Re = \sqrt{G} = \frac{\sqrt{|f|}\ L_0}{\nu}$$

by analogy with the Reynolds number UL/ν used in fluid mechanics, observing that in dimension 2, $U = \sqrt{|f|}$ has the dimension of a velocity ([5]).

We then have (see [30]).

Theorem 2.3. *The fractal (and Hausdorff) dimension of the universal attractor* $\tilde{X}(f)$ *is* $\leq c_0 G$ *where* c_0 *is a universal constant.*

An idea of the proof of theorem 2.3 is given is Sec. 5. The proof relies on a general result of estimate of dimensions of attractors in [5] and Lieb's inequality which generalizes the classical Sobolev inequalities.

Remarks 2.1.

i) A result of Babin-Vishik [2] shows that the Hausdorff dimension of $\tilde{X}(f)$ is bounded from below by $c\ G^{2/3}$, c a constant, proving that the upper bound $c_0 G$ is nearly optimal. The example in [2] corresponds to the space periodic case with an elongated period cell Ω, $L_1/L_2 = \varepsilon \ll 1$. For ε small enough it is shown in [2] the existence of unstable stationary solutions such that the number of unstable eigenvalues (and therefore the dimension

([5]) For a turbulent flow there is however no typical velocity and there is no physical evidence that $\sqrt{|f|}$ is of the order of a typical ("average") velocity.

of the corresponding unstable manifold which is included in $\tilde{X}(f)$) is $\geq c\, G^{2/3}$.

ii) Before theorem 2.3 several intermediate weaker results were proved. These results gave different bound for dim $\tilde{X}(f)$ in the periodic and bounded cases. In particular the method usedin the original proof of [11] gives (see [13]) :

Ω bounded : $\quad \dim \tilde{X} \leq c\, e^{Re}$

periodic case : $\dim \tilde{X} \leq c\, Re^4 \log Re$

The results of [2] and of Constantin-Foias [4] are :

Ω bounded $\quad \dim \tilde{X} \leq c\, Re^4$

periodic case : $\dim \tilde{X} \leq c\, Re^2 \log Re$

iii) Theorem 2.2 and 2.3 and the results below in dimension 3 are related to the fractal dimension of X and the Hausdorff dimension. The fractal dimension of a set is always larger than the Hausdorff dimension and in some sense takes into account self similarity of the set. There are examples of sets with a finite Hausdorff dimension and an infinite fractal dimension. There are also *denumberable* subsets of the interval [0,1] of $\mathbb{R}$ which has therefore 0 Hausdorff dimension and *a fractal dimension arbitrarily close to* 1.

This shows that the results concerning the fractal dimension may be much sharper than the result concerning the Hausdorff dimension.

3. THE THREE DIMENSIONAL CASE.

As indicated in the introduction, in the three dimensional case, the theory of existence and uniqueness of solutions for the initial value problem of the Navier-Stokes equations being still uncomplete, the results concerning the functional invariant sets are less precise. However we are able to give here an interesting physical interpretation to the results in connection with the (statistical) Kolmogorov theory of turbulence.

The functional form of the N.S.E. is still written as (1.3)(1.4). Given f in H (f is independant of t as in (2.1)) and given $u_o \in V = D(A^{1/2})$, the differential equation (1.3)(1.4) possesses a unique solution defined at least on a interval $[0,T(M)]$, where $T(M) = \dfrac{K}{(1+M^2)^2}$, $|A^{1/2}u_o| \le M$ and K depends on the data (other than u_o, i.e. f,ν,Ω). The solution u(.) is continuous with values in V on $[0,T(M)]$ and is analytic with values in D(A) on $]0,T(M)[$. We can thus define the mappings

$$S(t) : u_o \to u(t)$$

on the ball B(0,M) of V centered at 0 of radius M, for $t \in [0,T(M)]$. For $0 \le s,t \le s+t \le T(M)$, the operators S(.) enjoy the usual semi-group property $S(t).S(s) = S(t+s)$.

Functional Invariant Sets.

The definition of a functional invariant set X given in Sec. 2 has to be modified as follows : *This is a set* $X \subset V$, *bounded in* V , *such that*

i) $S(t)u_o$ *exists*, $\forall t > 0$, $\forall u_o \in X$ (6)

ii) $S(t)X = X$, $\forall t > 0$.

The definition given in Sec. 2 of an attractor applies as well but we do not know of the existence of any attractor in dimension 3, of course, nor

(6) For $u_o \in V$ and $t > 0$, the expression $S(t)u_o$ exists means that (1.3)(1.4) has a (unique) solution on $[0,t]$ which is bounded (and thus continuous) from $[0,t]$ with values in V.

the existence of an absorbing set and/or a universal attractor.

The results which are available concern solutions of (1.3)(1.4) which remain bounded in V and are not singular in the sense of Leray :

$$\sup_{t>0} |A^{1/2}u(t)| < \infty \tag{3.1}$$

If a solution of (1.3)(1.4) satisfies (3.1) then we can consider the set $X(f,u_o)$ defined as in (2.3). It is proved in [11] that this set is a functional invariant set for the N.S.E. which is bounded in V. *Also theorems* 2.1 *and* 2.2 *apply as well to any functional invariant set bounded in* V *in dimension* 3.

We are not able to give an analog of theorem 2.3 but we can give a physically interesting bound on the dimension of a functional invariant set.

<u>The Kolmogorov-Landau-Lifschitz estimate on the number of degrees of freedom of a turbulent flow.</u>

In [18], Landau-Lifschitz give an estimate, based on Kolmogorov theory of turbulence, of what they call the number N of degrees of freedom of a turbulent flow. The concept of degrees of freedom is not rigorously defined and their estimate which is based on averaging and dimensionality arguments is of the form

$$N \le c\left(\frac{L_o}{L_d}\right)^3 \tag{3.2}$$

where c is a constant, L_o is a typical macroscopic length (say as above the diameter of Ω), and L_d is the Kolmogorov dissipation length.

It follows from Kolmogorov theory that the eddies of size $\le L_d$ are damped exponentially, and therefore only the larger eddies have to be monitored for a proper description of a turbulent flow. Now since the number of cubes of edge L_d contained in Ω is of the order of $(L_o/L_d)^3$ and the number of eddies in the cube of size $\ge L_d$ is finite, we obtain the bound (3.2) for the number of parameters needed to monitor the large eddies and therefore the flow itself.

The definition of L_d in the Kolmogorov theory is

$$L_d = \left(\frac{\nu^3}{\varepsilon}\right)^{1/4} \qquad (3.3)$$

ν is as above and ε is the average dissipation rate of energy per unit volume and time ; it is a statistical (an ensemble average) of $\varepsilon(x,t) = \nu\,|\text{grad } u(x,t)|^2$, which is the local rate of dissipation of energy.

The result proved in [6] is precisely (3.2), *provided we interpret the number of degrees of freedom of the turbulent flow as the (fractal) dimension of the corresponding attractor* and provided we define L_d and ε with a proper averaging on the attractor X of $\varepsilon(x,t)$.

Given an attractor (or functional invariant set) X bounded in V, and given $p,q \in [1,\infty]$, we define $\varepsilon(p,q)$, the $\{p,q\}$-average of $\varepsilon(x,t)$ on X, as :

$$\varepsilon(p,q) = \limsup_{t\to\infty}\left\{ \sup_{u_0 \in X} \left(\int_0^t |\varepsilon(.,s)|^p_{L^q(\Omega;\frac{dx}{|\Omega|})} \frac{ds}{t}\right)^{1/p}\right\}$$

where $L^q(\Omega;\frac{dx}{|\Omega|})$ is the space of L^q functions on Ω for the averaged measure $\frac{dx}{|\Omega|}$ ($|\Omega|$ = the Lebesgue measure of Ω).

Setting $\varepsilon = \varepsilon(1,\infty)$, it was proved in Constantin-Foias-Mauley-T. [6] (see also [5]):

Theorem 3.1. *Let* X *be an attractor (a functional invariant set) bounded in* V. *Then its fractal dimension is bounded by* $c_1(L_0/L_d)^3$, *where* c_1 *is a universal constant and* L_0, L_d *are defined above.*

The proof which is based on the general results of [5] will be sketched in Sec. 4.

Remark 3.1. Using, as for theorem 2.3, a Lieb's inequality, we obtain the similar result with $\varepsilon = \varepsilon(1,\infty)$ replaced by $\varepsilon(1,5/4) \le \varepsilon(1,\infty)$, giving a sharper bound for the dimension. There is however no particular physical significance of $\varepsilon(1,5/4)$. The best bound would correspond to $\varepsilon = \varepsilon(1,1)$, but such a result seems to depend on the resolution (or at least a better understanding) of the initial value problem for the 3-dimensional N.S.E.

Numerical Analysis of the Navier-Stokes equations.

We conclude this Section with some comments on the significance of the above results for the numerical analysis of the Navier-Stokes equations.

The estimate (3.2) of the number of degrees of freedom of a turbulent flow has been used as an estimate on the number of parameters (or nodal values) which is necessary to describe a turbulent flow (see for instance S. Orszag [26]). Theorem 3.1 which proves (3.2) confirms the validity of this bound.

Given the number of parameters N, one can estimate the number of arithmetic operations per second which is necessary to solve numerically the N.S.E. equations *for real flows*, and this number is far beyond the possibilities of the present computers. Hopefully the supercomputers will get us closer to the necessary computing power and the estimate (3.2) is actually used to estimate the desirable computing power of the supercomputers.

Despite the important work already done ([7]) there is a long way to come for the numerical analysis of the Navier-Stokes equations in turbulent situations. Assuming the computing power is available, *we will be left with the important numerical problem of actually finding the* N *parameters which can describe the flow* since (3.2) and theorem 3.1 say nothing about the choice of the parameters. We will have also to find wether (3.2) is a theoretical bound ar a pratical one.

Let us mention some very partial results in that direction :

- In [9] the concept of determining modes was used : the parameters used to describe the flow are the Fourier series components. But in [9] the number of parameters which was proved to be necessary is much larger than the number of degrees of freedom (cGlogG for the space periodic case and c G^2 for the bounded case for a two dimensional flow, instead of cG as

([7]) See for instance A. Chorin [3], R. Glowinski [15], D. Gottlieb - S. Orszag [16], R. Peyret - R. Taylor [27], F. Thomasset [31], R. Temam [28].

given by theorem 2.3).

The question is then whether the estimates in [9] are not sharp enough, or the Fourier series parameters are not optimal, or there are no parameters giving exactly the bound cG of theorem 2.3.

- In [12] the concept of determining points was used : the values of the velocity vector on a discrete set $\mathcal{E}$ of points of Ω fully determine the large time behaviour of the flow (the "permanent regime"), provided each point of $\mathcal{E}$ is at a distance $< \alpha$ of some other point of $\mathcal{E}$, and α is sufficiently small. A natural conjecture is that α should be of the order of L_d (3-D case). Instead the results in [12] demand α to be of the order of $L_o \exp(L_o/L_d)$ which is much to high.

As above the question is whether the estimates in [12] can be improved, or a more delicate choice of the nodal points (say in finite differences) must be made or else, in the worse case, the number of parameters in (3.2) is a theoretical one.

These questions are important and some answer (theoretical or empirical) will be needed when the computations will start on supercomputers.

4. PRINCIPLE OF THE PROOFS.

We give here an idea of the proofs of the main theorems 2.3 and 3.1.

One of the ingredients of the proofs is the concepts of uniform Lyapunov exponents and a general result in [5], and another one is Lieb's inequality.

Uniform Lyapunov Exponents.

Let H be a Hilbert space, $X \subset H$ a compact subset and S a continuous mapping from X into X. In the application X will be a functional invariant set and $S = S(t)$ for some $t > 0$.

It is assumed than S is "uniformly differentiable on X" which means that for every $u_o \in X$, there exists $L(u_o)$ linear continuous in H, such that

$$\sup_{|v_o - u_o|_H \le \varepsilon} \frac{|S(v_o)-S(u_o)-L(u_o)(v_o-u_o)|_H}{|v_o-u_o|_H} \to 0$$

as $\varepsilon \to 0$ ($\varepsilon > 0$). Note that $L(u_o)$ need not to be unique. In the applications (N.S.E.), $S(t) = S$ is differentiable in the classical sense if the dimension n=2 and is differentiable in the weak sense above if n=3 (see [5]) ; we write $L(u_o,t)$ for $L(u_o)$ when $S = S(t)$.

Given a linear compact operator L in H, we denote by $\alpha_1(L) \geq \alpha_2(L) \geq \ldots \geq 0$, the eigenvalues of the self-adjoint positive compact operator $(L^*L)^{1/2}$ in decreasing order ; they are also the axes of the ellipsoid $LB_H(0,1)$, the image by L of the unit ball in H (centered at 0). We also set $\omega_m(L) = \alpha_1(L) \ldots \alpha_m(L)$. It can be proved that $\omega_m(L)$ is the norm in the exterior product $\Lambda^m H$ of $\Lambda^m L$. This number indicates how L changes the volume in dimension m :

$$\omega_m(L) = \text{Sup meas}_m(LB_m)$$

where the supremum is for all the unit balls in subspaces of dimension m of H, and $\text{meas}_m(LB_m)$ is the m-dimensional measure of the image LB_m (an m-dimensional ellipsoid).

Now let $L = L(u_o,t)$, $u_o \in X$, $t > 0$, X a functional invariant set for the N.S.E. Let also

$$\omega_m(t) = \sup_{u_o \in X} \omega_m(L(u_o,t)).$$

It is easily seen that the functions $t \to \omega_m(t)$ are sub-exponential , i.e.

$$\omega_m(t+s) \leq \omega_m(t).\omega_m(s),$$

and because of that, $\omega_m(t)^{1/t}$ tends to a limit π_m as $t \to \infty$. The *uniform Lyapunov numbers* $\mu_1, \mu_2, \ldots,$ on X, are then recursively defined by

$$\mu_1 = \log \pi_1, \quad \mu_1 + \ldots + \mu_m = \log \pi_m, \quad m \geq 2,$$

i.e. $$\mu_m = \log(\pi_m/\pi_{m-1}), \quad m \geq 2.$$

Under the above hypotheses (and some further technical one which we do not state) it is proved in Constantin-Foias-T. [5], that if SX = X and

$$\mu_1 + \ldots + \mu_m < 0 \tag{4.1}$$

for some m, then

- the Hausdorff dimension of X, $\dim_H(X)$ is $\leq m$
- the fractal dimension of X, $\dim_F(X)$ is $\leq$ some expression which can not be described without introducing further quantities but happens to be $\leq c\, m$, in the applications, c a constant.

We are then left with the problem of finding m such that (4.1) holds.

The first variation equations.

Let $u(.) = S(.)u_0$ be a solution of (1.3)(1.4). Then the (generalized) differential operator $L(u_0,t)$ (the differential of $u_0 \to S(t)u_0$) is defined through the first variation equations which are a formal linearisation of (1.3) :

$$\frac{dU}{dt} + \nu AU + B(u,U) + B(U,u) = 0 \tag{4.2}$$

$$U(0) = \xi \tag{4.3}$$

It is proved in [5] that (4.2)(4.3) possesses a unique solution U for every $\xi \in H$, for every $t > 0$ and that $L(u_0,t).\xi$ is precisely $U(t)$.

We then have

$$\omega_m(L(u_0,t)) = \sup_{\substack{\xi_1,\ldots,\xi_m \in H \\ |\xi_j| \leq 1,\ \forall j}} |U_1(t) \wedge \ldots \wedge U_n(t))|_{\Lambda^m H}$$

where $\xi_1, \ldots, \xi_m \in H$ and $U_1, \ldots, U_m$ are the corresponding solutions of (3.4) (3.5). Then it is proved that

$$\mu_1 + \ldots + \mu_m \leq \limsup_{t\to\infty} \sup_{u_0 \in X} \left(-\frac{1}{t}\int_0^t \sup_{Q_m}(\mathrm{Tr}\mathcal{A}(u) \circ Q_m)ds\right))$$

where $Q_m = Q_m(s)$ is the projection in H on the space spanned by $U_1(s), \ldots, U_m(s)$ and $\mathcal{A}(u) = \mathcal{A}(u(s))$ is the linear mapping form D(A) into H :

$$U \to \nu AU + B(u(s),U) + B(U,u(s)).$$

The problem of finding m such that (3.3) holds in then reduced to that of properly estimating the trace of $\mathcal{A}(u) \circ Q_m$. For theorem 3.1 this is done by a direct analysis ; for theorem 2.3 this relies on an appropriate extension of Lieb's inequality which we will recall.

The trace of $\mathcal{A}(u(s)).Q_m(s)$ is computed as follows : let $\phi_j = \phi_j(s)$ be an orthonormal basis of H such that $\phi_1(s), \ldots, \phi_m(s)$ is an orthonormal basis of $Q_m(s)H$ (thus $\phi_j(s) \in D(A)$). Then

$$- \mathrm{Tr}\mathcal{A}(u(s)) \circ Q_m(s) = - \sum_{j=1}^{\infty} (\mathcal{A}(u(s)) \circ Q_m(s)\phi_j(s), \phi_j(s))$$

$$= - \sum_{j=1}^{m} (\mathcal{A}(u(s))\phi_j(s), \phi_j(s))$$

$$= \text{(since } (B(\phi,\psi),\psi) = 0,\ \phi,\psi \in V)$$

$$= - \sum_{j=1}^{m} |A^{1/2}\phi_j|^2 - \sum_{j=1}^{m} (B(\phi_j,u),\phi_j).$$

We then have to estimate $\sum_{j=1}^{m}(B(\phi_j,u),\phi_j)$ in term of $\sum_{j=1}^{m}|A^{1/2}\phi_j|^2$ and use the fact that $\sum_{j=1}^{m}|A^{1/2}\phi_j|^2 = \lambda_1 + \ldots + \lambda_m$, where the λ_j's are the eigenvalues of $A(0 \leq \lambda_1 \leq \lambda_2 \ldots)$, and the behaviour of λ_m as $m \to \infty$ is well known (see[25]).

As mentioned above the estimate of $\sum_j (B(\phi_j,u),\phi_j)$ in term of $\sum_j |A^{1/2}\phi_j|^2$

relies on Sobolev's inequality for theorem 3.1 and on a generalization of Lieb's inequality for theorem 2.3.

Lieb-Thirring Inequality

We only state the original Lieb-Thirring inequality [22] for $H_0^1(\Omega)$, a bounded open set in $\mathbb{R}^2$. The extension which is necessary for our application will be given elsewhere.

For Ω a bounded set of $\mathbb{R}^2$, it is well known that $H_0^1(\Omega) \subset L^p(\Omega)$, $\forall p < \infty$ and, for $p = 4$, the Sobolev inbedding follows from the Gagliardo-Nirenberg inequality :

$$\int_\Omega \phi^4(x)\,dx \le c\left(\int_\Omega |\nabla\phi(x)|^2 dx\right)\left(\int_\Omega \phi^2(x)\,dx\right), \quad \forall\phi \in H_0^1(\Omega) \tag{4.4}$$

Now let ϕ_j be an orthonormal basis of $L^2(\Omega)$ with $\phi_j \in H_0^1(\Omega)$, $\forall j$. We infer from (3.6) that

$$\begin{cases} \int_\Omega \phi_j^4 dx \le c \int_\Omega |\nabla\phi_j|^2 dx, & \forall j \\ \sum_{j=1}^m \int_\Omega \phi_j^4 dx \le c \sum_{j=1}^m \int_\Omega |\nabla\phi_j|^2 dx, & \forall m \\ \int_\Omega \left(\sum_{j=1}^m \phi_j^2\right)^2 dx \le c(m) \sum_{j=1}^m \int_\Omega |\nabla\phi_j|^2 dx, & \forall m, \end{cases} \tag{4.5}$$

where $c(m)$ depends on m and $\to +\infty$ as $m \to \infty$.

Now one of the inequalities in [22] is precisely (4.5) with $c(m)$ replaced by K_1, K_1-*independant* of m.

5. OTHER PROBLEMS.

We give some indications on similar results for two other problems :

- the flow in a bounded domain Ω of $\mathbb{R}^2$ or $\mathbb{R}^3$ with a non homogeneous boundary condition for u on Γ
- the Benard convection problem.

The Non Homogeneous Navier-Stokes Equations.

For simplicity we restrict ourselves to the two dimensional case : Ω is a bounded open set of $\mathbb{R}^2$ with a smooth boundary Γ. We consider the equations (1.1)(1.2) for $x \in \Omega$ and the boundary condition

$$u = \phi \text{ on } \Gamma \tag{5.1}$$

where ϕ is given. We assume as in (2.1) that ϕ is independent of time and we assume that ϕ is extended inside Ω as a function Φ in $H^2(\Omega)^2$ satisfying $\text{div } \Phi = 0$.

For the solution of (1.1)(1.2)(5.1) it is convenient to make a change of variable and set $u = v + \Phi$. Then the equation for v can be written in a functional form similar to (1.3)(1.4)

$$\frac{dv}{dt} + \nu Av + B(v,\Phi) + B(\Phi,v) + B(v,v) = g \tag{5.2}$$

$$v(0) = v_0 \; (=u_0 - \Phi) \tag{5.3}$$

with $g = - \nu A\Phi - B(\Phi,\Phi) + f$.

The existence of an absorbing set and of a universal attractor are easy to prove. Theorems 2.1 and 2.2 can be extended to this case as well. However a difficulty arises when we try to extend theorem 2.3 to this case: consider a Reynolds number $Re = UL_0/\nu$ where the "typical velocity U" is some average of ϕ. Trying to estimate the dimension of the universal attractor in term of Re one would expect a bound of $\dim \tilde{X}$ of the order of Re or at most a power of Re. The bounds found at the moment are of the form exp(Re) and a reason for that is the following : we use for the extension Φ of ϕ into Ω, a technique due to E. Hopf (see for instance[23][28]) and the H^1-norm of this Φ is of the order of $\exp(\frac{1}{\nu})$, giving an absorbing set with a diameter of the same order. The obtention of better estimates is an open problem.

The Benard Convection Problem

We consider the Benard convection problem for an horizontal of fluid $0 < x_2 < 1$. The equations are the Navier-Stokes equations couple with the heat equation. The equations in the non-dimensional case take the form :

$$\frac{\partial u}{\partial t} + (u.\nabla)u - \nu\Delta u + \nabla p = e_2\theta \tag{5.4}$$

$$\nabla u = 0 \tag{5.5}$$

$$\frac{\partial \theta}{\partial t} + (u.\nabla)\theta - u_2 - K\Delta\theta = 0 \tag{5.6}$$

where e_2 is the unit vector of the x_2 axis and, now, (in non-dimensional form) $\nu = \frac{1}{\sqrt{r}}$, $K = \frac{1}{\sqrt{PrRa}}$, Gr = the Grashof number, Pr = the Prandtl number, Ra = the Rayleigh number.

The boundary condition are the space periodicity with period in the $_1$ direction and $u = 0$ at $x_2 = 0$ and $x_2 = 1$ and $\theta = T - T_0 - x_2(T_1 - T_0)$ vanishes at $x_2 = 0$ and $x_2 = 1$.

The existence of a universal attractor with a finite fractal dimension can be proved. It was conjectured by O. Mauley that the number of degrees of freedom of the flow should be of the order of G and a partial result in [9] concerning determining modes gave a bound on dimension of the order $G\log G$.

In an article to appear [8] it is now proved that the dimension of the universal attractor is bounded by $c\, Gr(1+Pr^{3/2})$. The correction term involving Pr is not harmful since Pr is not usually large. The proof of this bound uses again the general results of Constantin-Foias-Temam [b] and a further extension of Lieb-Thirring inequality.

Remark 5.1. An interesting property which is actually essential for the derivation of the bounds on the solution is that $-1 < \theta < 1$ on the universal attractor $\tilde{X}$, which corresponds to a maximum prin ple property, $T_1 < T < T_o$:

at any point of X the temperature T lies between the temperatures of the lower (T_0) and upper planes (T_1).

REFERENCES.

[1] V. Arnold, *Equations differentielles ordinaires*, ed. MIR, Moscou.

[2] A.V. Babin, M.I. Vishik, *Les attracteurs des équations d'évolution aux dérivées partielles et les estimations de leur dimensions*, Usp. Math. Nauk, 38, 4 (232), 1983, p. 133-187.

[3] A. Chorin, *Lectures on Turbulence Theory*, Publish on Perish Inc, Baston, 1975.

[4] P. Constantin, C. Foias, *Global Lyapunov Exponents, Kaplan Yorke formulas and the Dimension of the Attractors for 2 D Navier-Stokes Equations*, Comm. Pure Appl. Math., to appear.

[5] P. Constantin, C. Foias, R. Temam, *Attractors representing turbulent flows*, Memoirs of AVIS, 1985.

[6] P. Constantin, C. Foias, O.P. Manley, R. Temam, *Connexion entre la theorie mathématique des équations de Navier-Stokes et la théorie conventionnelle de la turbulence*, C.R. Ac. Sci. Paris, Série I, 297, 1983, p. 599-602, and *Determining Modes and fractal Dimension of turbulent Flows*, J. Fluid. Mech.. to appear.

[7] H. Federer, *Geometric Measure Theory*, Springer-Verlag, Berlin-Heidelberg-New York, 1969.

[8] C. Foias, O. Manley, R. Temam, *Attractors for the Benard Problem : Existence and physical bounds of their fractal dimension, to appear.*

[9] C. Foias, O. Manley, R. Temam, Y. Trève, *Asymptotic Analysis of the Navier-Stokes Equations*, Physica D., Vol. 6D, 1983.

[10] C. Foias, G. Prodi, *Sur le comportement global des solutions non stationnaires des équations de Navier-Stokes en dimension 2*, Rend. Sem. Mat. Padova, 39, 1967, p. 1-34.

[11] C. Foias, R. Temam, *Some analytic and geometric Properties of the Solutions of the Navier-Stokes Equations*, J. Math. Pures

Appl., 58, 1979, p. 339-368.

[12] C. Foias, R. Temam, *Determination of the solutions of the Navier-Stokes Equations by a Set of nodal Values*, Math. Comput., 43, n°167, (1984), p. 117-133.

[13] C. Foias, R. Temam, *On the Hausdorff Dimension of an Attractor for the two-dimensional Navier-Stokes Equations*, Phys. Lett., 93A, n°9, 1983, p. 451-454.

[14] U. Frisch, *Chaotic behaviour in Deterministic Systems*, Les Houches, 1983, North-Holland, Amsterdam, 1984.

[15] R. Glowinski, *Numerical method for nonlinear variational problems*, Springer , 1984.

[16] D. Gott ieb, S. Orszag, *Numerical Analysis of spectral methods*, SIAM Publ., 1977.

[17] C. Guillopé, *Comportement à l'infini des solutions des équations de Navier-Stokes et propriété des ensembles fonctionnels invariants (ou attracteurs)*, Ann. Inst. Fourier (Grenoble), 32, 3, 1982, p. 1-37.

[18] L. Landau, I.M. Lifschitz, *Fluid Mechanics*, Addison-Wesley, New York, 1953.

[19] J. Leray, *Etude de diverses équations intégrales non linéaires et de quelques problèmes que pose l'hydrodynamique*, J. Math. Pures Appl., 12, 1933, p. 1-82.

[20] J. Leray, *Essai sur les mouvements plans d'un liquide visqueux que limitent des parois*, J. Math. Pures Appl., 13, 1934, p. 331-418.

[21] J. Leray, *Essai sur les mouvements plans d'un liquide visqueux emplissant l'espace*, Acta Math., 63, 1934, p. 193-248.

[22] E. Lieb, W. Thirring, *Inequalities for the Moments of the eigenvalues of the Schroedinger Equations and their relation to Sobolev Inequalities*, p. 269-303, in *Studies in Mathematical Physics : Essays in Honor of Valentine Bargman*, E. Lieb, B. Simon, A.S. Wightman Eds, Princeton University Press, Princeton, N.J. 1976.

[23] J.L. Lions, *Quelques méthodes de résolution des problèmes aux limites non linéaires*, Dunod, Paris, 1969.

[24] B. Mandelbrot, *Fractals : Form, Chance and Dimension*, Freeman, San Francisco, 1977.

[25] G. Métivier, *Valeurs propres d'opérateurs définis sur la restricti de systèmes variationnels à des sous-espaces*, J. Math. Pures Appl., 57, 1978, p. 133-156.

[26] S. Orszag, Lecture at the VII the International Congress on Mathematical Physics, Boulder, Colorado , 1983, Physica 124 A 1984.

[27] R. Peyret, R. Taylor, *Computational methods for fluid flows*, Sprin Verlag, 1982.

[28] R. Temam, *Navier-Stokes Equations, Theory and Numerical Analysis*, 3 rd Ed, North-Holland, Amsterdam, 1984.

[29] R. Temam, *Navier-Stokes Equations and Nonlinear Functional Analysi* CBMS-NSF Regional Conference series in Applied Mathematics, SIAM, Philadelphia, 1983.

[30] R. Temam, *Infinite dimensional dynamical systems in fluid mechanic* in *Nonlinear functional Analysis and Applications*, Proccedings of the AMS-Summer Research Institute, Berkeley, 1983, F. Brow Ed..

[31] F. Thomasset, Implementation of Finite Element methods for Navier-Stokes Equations, Springer Verlag, 1981.

[32] A.V. Babin, M.I. Vishik, *Attracteurs maximaux dans les équations au dérivées partielles*. In thits volume.

R. TEMAM
Université de Paris-Sud et CNRS
Laboratoire d'Analyse Numérique
Bat. 425
91405 ORSAY CEDEX
FRANCE